Hydrologic Hazards Science at the U.S. Geological Survey

Committee on U.S. Geological Survey Water Resources Research

Water Science and Technology Board

Commission on Geosciences, Environment, and Resources

National Research Council

NATIONAL ACADEMY PRESS
Washington, D.C. 1999

Support for this project was provided by the U.S. Geological Survey under Grant No. 1434-93-A-0982.

International Standard Book Number 0-309-06282-9

Copies of this report are available from the Water Science and Technology Board, National Research Council, 2101 Constitution Avenue, N.W., Washington, DC 20418.

Cover by Van Nguyen, National Academy Press, using photos provided by U.S. Geological Survey.

Printed in the United States of America

v

Preface

This report is a product of the Committee on U.S. Geological Survey (USGS) Water Resources Research, which provides consensus advice to the Water Resources Division (WRD) of the USGS on scientific, research, and programmatic issues. The committee is one of the groups that works under the auspices of the Water Science and Technology Board of the National Research Council. The committee considers a variety of topics that are important scientifically and programmatically to the USGS and the nation and issues reports when appropriate.

This report concerns the work of the WRD in science and technology relevant to hydrologic hazards, which include droughts, flooding, and related phenomena, such as debris flows. Natural hazards-related work can be found in programs throughout the USGS; some of the activities (e.g., debris flow assessment or hazards communication efforts) offer excellent opportunities for multidisciplinary, cross-divisional, and interagency collaboration. Within the WRD, hydrologic hazards-related work is dispersed throughout, providing some of the most challenging subject matter for research, interpretive studies, and data collection efforts.

Losses of life and property due to hydrologic hazards in the United States are large—over $2.5 billion per year in direct damages and 100 lives annually due to floods alone. While the nation has struggled with various approaches to control river flows or avoid their ravages, these damages and losses continue to increase. Construction of hydraulic projects for water control was the national choice for decades until 30 years ago, when alternative insurance-based land management and "mitigation" activities began to become an accepted strategy to reduce human and property exposure. In either case, sophisticated understanding of hydrologic and hydraulic processes, data essential to hydrologic analyses, and the development and dissemination of information provide the underlying bases for policy and management. Streamflow data collected by the USGS for over 100 years and the modern water science and technology carried out by the WRD form the cornerstone for national, regional, and local efforts to cope with hydrologic hazards by providing continued, up-to-date information about water conditions and understanding of hydrologic phenomena.

This report attempts to help shape and improve the overall framework for the USGS's efforts relevant to hydrologic hazards. The report addresses hydrologic research needs; issues associated with the all-important stream gaging network; assessment and interpretation methods; and coordination, dissemination, and outreach activities. A short report such as this, prepared by an outside group such as ours, cannot provide an in-depth assessment of all germane WRD programs and projects but instead is a more general document intended to provide strategic advice to WRD management.

The committee began its review in late 1996, when it laid out plans for the study. Subsequently, the committee met four times before completing this report. At meetings, members were briefed by USGS personnel on a variety of programs and activities. The committee learned about WRD's efforts relevant to hydrologic hazards in many hydrologic regions. Briefings on and/or visits to such diverse issues and sites as the 1988-1992 North Dakota drought, flood hazard information dissemination in Louisiana, Mt. Rainier flow hazards, coordinated water management in the Pacific Northwest, the floods of 1997 on the Red River of the North, and research in such topics as discharge frequency analysis, hydroclimatology, bridge scour, and new hydrologic technologies provided invaluable information for review and assessment. Committee members drafted individual contributions and deliberated as a group to achieve consensus on the content of this report. We hope that by maintaining a broad, forward-looking perspective, our assessment will prove useful.

As the study proceeded and the committee became more cognizant of USGS activities, productive discussions occurred among committee members and personnel from the USGS and other organizations. This interaction was critical to the success of the project. The committee heard from more than 20 USGS staff members and several individuals from the National Weather Service, U.S. Army Corps of Engineers, and other organizations. The list of individuals providing information to the committee is too long to include in this preface, but we are indebted for the many perspectives and for the information provided. We do wish to single out three individuals from the USGS who interacted throughout the project and thank them for the assistance, information, and cooperation they provided: Thomas H. Yorke, Jr., chief of the Office of Surface Water; Robert M. Hirsch, chief hydrologist; and Gail E. Mallard, hydrologist, who serves as USGS's continuing liaison with our committee. We also wish to acknowledge the NRC's Board on Natural Disasters (BOND, a "sister" unit of the Water Science and

Technology Board), which provided some assistance with this project. Notably, BOND members James (Jeff) F. Kimpel participated in some of our deliberations and contributed written input, and Frank H. Thomas participated in the review (see below).

The committee hopes that this report will help promote development and appreciation for improved hydrologic data, information, and knowledge as it supports the nation's efforts to reduce the tolls of hydrologic hazards. Continued, significant, and successful contributions to this area of hydrologic science by the USGS are essential.

This report has been reviewed by individuals chosen for their diverse perspectives and technical expertise, in accordance with procedures approved by the National Research Council's (NRC) Report Review Committee. The purpose of this independent review is to provide candid and critical comments that will assist the authors and the NRC in making the published report as sound as possible and to ensure that the report meets institutional standards for objectivity, evidence, and responsiveness to the study charge. The content of the review comments and draft manuscripts remain confidential to protect the integrity of the deliberative process. We thank the following individuals for their participation in the review of this report: S. "Rocky" Durrans, University of Alabama; George M. Hornberger, University of Virginia; L. Douglas James, National Science Foundation; Rebecca T. Parkin, The George Washington University; Marc Parlange, The Johns Hopkins University; and Frank H. Thomas, consultant.

While the individuals listed above provided many constructive comments and suggestions, responsibility for the final content of this report rests with the authoring committee and the NRC.

Kenneth R. Bradbury
Chairman, Committee on USGS
Water Resources Research

Contents

Executive Summary

Losses of life and property in the United States—and throughout the world—resulting from hydrologic hazards, including floods, droughts, and related phenomena, are significant and increasing. Public awareness of, and federal attention to, natural disaster reduction, with a focus on mitigation or preparedness so as to minimize the impacts of such events, have probably never been greater than at present. With over three-quarters of federal disaster declarations resulting from water-related events, national interest in having the best-possible hydrologic data, information, and knowledge as the basis for assessment and reduction of risks from hydrologic hazards is clear.

The U.S. Geological Survey (USGS) plays a variety of unique and critical roles relevant to hydrologic hazard understanding, preparedness, and response. The agency's data collection, research, techniques development, and interpretive studies provide the essential bases for national, state, and local hydrologic hazard risk assessment and reduction efforts. This work includes some of the more traditional activities of the Water Resources Division (e.g., streamflow measurement) and some of the more innovative interdisciplinary activities (e.g., hydrologic research, educational outreach, real-time data transmission, and risk communication) being pursued in cooperation with other divisions of the USGS, other federal and state agencies, and other local entities. This report aims to help shape a strategy and improve the overall framework of USGS efforts in these important areas.

The USGS is well known as the nation's primary supplier of reliable streamflow and water-level data and this role is essential. But the USGS should also expand its efforts to document and analyze extreme hydrologic events, both during and after their occurrence. The agency is ideally positioned to collect and archive the critical hydrologic information necessary to improve our understanding of how and why such extreme events happen and to improve our ability to predict them. Specifically, this scientific work should proceed according to a strategy that features:

- Maintaining the integrity and continuity of the national stream gaging network;
- Improved stream gaging network design, measurement techniques, and instrumentation for the measurement of streamflow and stream stage;
- Postaudits of the technical response and prediction of major floods;
- Improved discharge measurements of extreme floods;
- Improved approaches for regional flood-frequency estimation;
- Improved methods for drought forecasting;
- Investigations of the long-term stationarity of floods and droughts; and
- Improved techniques for low flow frequency analysis, and its relevance to instream flow management and ecologically based regulatory criteria.

The USGS should build on its experience in managing and disseminating water resources data as a critical part of the hydrologic hazards program. In particular, the USGS should place new emphasis on rapid data acquisition and retrieval during extreme events and explore new methods for integrating datasets over several scientific disciplines. Geographic information systems technology may offer techniques for integrating, analyzing, and displaying dissimilar datasets for improved analyses of hydrologic hazards.

Rapid expansion of Internet use has had a great influence on USGS's approach to disseminating hydrologic data and related information. The agency is currently offering real-time data on the Internet for more than 3,900 stream gaging stations, and the number will continue to grow. This capability of acquiring and disseminating data in real time expands the "customer-base" and "products" of the USGS. The principal customers are no longer only researchers, planners, and designers; customers now include emergency managers and the public. Beyond the expansion of real-time monitoring networks, the USGS is encouraged to add risk-based interpretation to its hydrologic data, such as comparison with historical data and simulated visualizations of flood inundation areas. USGS can take the lead in improving hydrologic understanding through improved "visualization" approaches that integrate the agency's expertise in long-term monitoring, mapping, and process modeling. The committee recommends that the USGS consider giving significant new attention to outreach activities.

This report concludes that the USGS should play a prominent role in risk-based decision making with respect to hydrologic hazards. Specifically, in addition to being a provider of data, the USGS should conduct research on techniques for estimating the probability and magnitude of extreme hydrologic events in the context of risk-based decision making. This work should consider how changes in land use, climate, and streamflow regulation influence hydrologic hazards. It should also improve integrated risk and process models related to floods and droughts. The USGS should couple its role in the analysis of risk to its outreach role in communicating to the general public what that risk means. One way to characterize this outreach mission would be to describe the USGS role as helping decision makers avoid being "surprised."

The ultimate goal of the hydrologic hazards program is to assist in protecting the lives and property of citizens from naturally occurring hazards while at the same time maintaining and protecting ecological communities. This goal requires that hazards information and research results be communicated to the public, and to public officials, in a timely and understandable manner. It is critical that the USGS maintain and develop liaisons with federal and nonfederal institutions in the research, management, and user communities to assure that efforts are pusued in an integrated and coordinated fashion. In addition, USGS scientists should be encouraged to participate as individuals in public discussions of hazards issues.

1

Introduction

Hydrologic hazards of various types present myriad technical and public policy challenges in the United States and worldwide and are defined as extreme events associated with water occurrence, movement, and distribution. Specifically, hydrologic hazards include flooding and related events (e.g., landslides and river scour and deposition) and droughts; coastal flooding and related phenomena are not included. In the United States, about 7 percent of the land area is subject to flooding, about one-third of the nation's streams experience severe erosion problems, landslides and mud slides are commonplace in some areas, and virtually all of the nation is susceptible to drought. Floods are generally regarded as the most significant of hydrologic hazards, with losses amounting to over $2.5 billion in direct damages and nearly 100 lives lost annually (FIFMTF, 1992). Hydrologic hazards can have sudden adverse effects on many people, including threats to public safety and costly interference with commerce. There is a continuing need for monitoring, regional studies, and research on hydrologic hazards in order to better understand physical processes, inform improved decision making, and ultimately help lessen the impacts of hydrologic hazards.

Hydrologic hazards are the focus of important activities carried out by the U.S. Geological Survey (USGS), including:

• *Monitoring* of streamflows to support the efforts of other organizations engaged in operations such as flood forecasting and reservoir management and the provision of information during emergency conditions.

• *Interpretive studies* to improve understanding of the hydrologic characteristics of floods and droughts.

• *Research and technique development* concerning topics such as frequency analysis of extreme events to support planning of hydraulic structures and flood-plain delineation and management, hydrologic modeling, paleoflood and hydro-climatic investigations, and scour analyses.

Much of the work performed by the USGS Water Resources Division has focused on developing and providing information to help minimize the uncertainties and lessen impacts associated with hydrologic hazards. Monitoring activities, in particular, provide the foundation for significant hydrologic hazards-related work performed by other agencies, including planning, forecasting, and emergency response.

Given the importance of hydrologic hazard information, the USGS seeks to continually improve its approaches, techniques, and products. The agency must also ensure that its work is focused on the most significant issues and that it provides information to address contemporary needs of other agencies and the public. In that regard, the objective of this report is to provide strategic guidance to the USGS on its hydrologic hazards research program. A number of broad questions stimulated and shaped the committee's deliberations that form the basis for this report. These questions, though not all of equal importance, are grouped below according to similar types of activity:

Aerial view of flooding in Grand Forks, ND, during spring of 1997. Photo courtesy of U.S. Geological Survey.

• What are the flood and drought information needs of USGS cooperators and others? What constitutes an adequate national stream gaging network for hydrologic hazard analyses and warnings? What should the USGS be doing to apply hydroclimatic understanding to produce improved hydrologic hazard information? How can the USGS account for multiple hazard events, particularly regional characterization of risk?

• What are important national needs for hydrologic hazard research by the USGS? For example, are current approaches for estimation of flood frequencies and for characterizing drought in need of improvement? What should be the focus of the agency's programs related to river scour, deposition, and landslides? What are the water quality and ecosystem implications of hydrologic hazards?

• Are there opportunities for the USGS to improve the interface with other agencies on hydrologic hazard research? Are there opportunities to improve the transfer of technology between USGS and the water management and planning entities? To what extent should the USGS be involved in communicating information about probability and consequences of natural hazards to the public?

The remainder of this report contains the analyses, conclusions, and recommendations of the study. Chapter 2 provides broad and historical context for USGS hydrologic hazards work, and succeeding chapters (3 and 4) address matters related to data collection, techniques development, research, and interpretive studies. Because of its increasing importance to the USGS, the communication of information on hydrologic hazards is featured in Chapter 5. Finally, Chapter 6 summarizes conclusions and recommendations that stem from the study.

As previously indicated, this report is intended to provide strategic guidance to the USGS for its activities related to hydrologic hazards. As such, it is not a comprehensive interagency science plan for all work related to hydrologic hazards. Nonetheless, since the USGS does not operate in isolation and indeed carries out much of its work in support of or in cooperation with others, it is expected that this report should be valuable to scientists and managers in other agencies and organizations interested in hydrologic hazards science and information.

2

Water Management and Information Needs

Floods, droughts, and other natural disasters have drawn special attention in both religious and secular literature from the beginning of time, and the modern world seems almost as vulnerable to the whims of nature as ever before. Despite massive investments in dams and drainage and other facilities to reduce the risks associated with these events, economic and environmental damages resulting from them continue to increase. Although the data are admittedly subject to uncertainty, especially prior to 1950, Burton et al. (1993) cite several sources that indicate that worldwide deaths from natural disasters are substantially lower now than in the first half of this century. While the number of deaths has decreased, the number of incidents with at least 100 deaths has sharply increased, possibly resulting from the increased concentration of populations in urban areas. Estimates are that economic losses also are increasing rapidly. Some of those trends may be explained simply by better reporting, but factors such as urbanization and increasing concentration of the population in coastal areas have no doubt contributed to increased risk. Of the reported incidents causing at least 100 deaths, floods are the leading cause (40 percent), and the other primary hydrologic hazard, drought, accounts for an additional 15 percent. Tropical cyclones and earthquakes account for 35 percent. Subsequent sections of this chapter include an overview of the nature and magnitude of damages caused by floods and droughts and an examination of organizational arrangements for addressing floods and droughts in the United States, with particular attention to the roles of various government management agencies. A special section is devoted to the role of the U.S. Geological Survey (USGS). The concluding section identifies key issues and their research and data implications.

FLOODS

Flooding is the most widespread hydrologic hazard in the United States and throughout the world. A recent compilation of data on events leading to declarations of disasters under the 1988 Stafford Disaster Relief and Emergency Assistance Act indicates that of the 295 declarations during the period December 1988 to May 1996, one-third were due to flooding (Godschalk et al., 1997). Another 11 percent were due to a combination of tornadoes and flooding. Those numbers do not include coastal storms and hurricanes, which cause a combination of wind and water damage. Because damages from a small number of devastating earthquakes and hurricanes account for a disproportionate share of all relief provided under the Stafford Act since 1988, riverine flooding accounted for a lesser but important share of the total, namely 17 percent. Not only is flooding the most frequently occurring disaster, but it is also the most widespread. Flood disaster declarations under the Stafford Act have been made in 42 states.

Because the Stafford Act broadened the range of disasters for which relief is made available, the proportion of all types of disasters attributable to flooding decreased after 1988. The Federal Interagency Floodplain Management Task Force (FIFMTF), citing data from the Federal Emergency Management Agency,

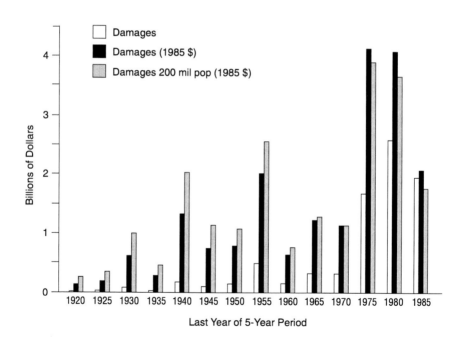

FIGURE 2.1 Average annual flood damages over five-year periods. Source: FIFMTF (1992).

Flood damage to bridge and road. Note person on bridge for scale. Photo courtesy of U.S. Geological Survey.

indicated that floods and hurricanes accounted for three-fourths of all presidentially declared disasters from 1965 through 1989 (FIFMTF, 1992).

The FIFMTF also reported that over the 70-year period from 1916 to 1985 there were, on average, 101 flood-related deaths annually in the United States and that there was no indication that the number of deaths per capita was changing (FIFMTF, 1992). While no trend was discernible in loss of life, a definite increase in economic losses over that period was noted. Those damages, adjusted for inflation and population increase, are shown in Figure 2.1. Average annual damage in the 1970s was in the neighborhood of $4 billion per year in 1985 dollars. Although that average dropped to just over $2 billion in 1980-1985, the Interagency Floodplain Management Review Committee estimated that the Midwest floods in 1993 caused $12 billion to $16 billion in damages (IFMRC, 1994).

FLOOD MANAGEMENT STRATEGIES

Strategies

The FIFMTF formulated a well-developed framework for floodplain management, consisting of four complementary strategies:

- Modify susceptibility to flood damage and disruption through floodplain regulations, development and redevelopment policies, disaster preparedness, disaster assistance, floodproofing and flood forecasting, warning systems, and emergency plans.

• Modify flooding by constructing dams, reservoirs, levees, floodwalls, channel alterations, high-flow diversions, land treatment, and onsite detention.

• Modify the impact of flooding on individuals and communities through information and education, flood insurance, tax adjustments, and postflood recovery.

• Restore and preserve natural and cultural floodplain resources.

Implementation of these strategies takes place through a complex array of federal, state, and local government and private-sector activities. Federal agencies with prominent roles in managing flood risk are the Federal Emergency Management Agency (FEMA), the U.S. Army Corps of Engineers (USACE), the Natural Resources Conservation Service, the Federal Crop Insurance Corporation, other agencies of the U.S. Department of Agriculture, and the National Weather Service. At times, the web of institutional arrangements and specific policies becomes so complex that it appears to be fragmented, and it is difficult to envision how well the broad policy framework formulated by FIFMTF is being implemented. It is in the details of these institutional arrangements and policies where the strategy's success or failure may rest. Given the bottom line of a rising rate of economic losses related to floods, it appears that there are serious weaknesses in some aspects of these policies.

Structural Measures

From the mid-1930s until the decade of the 1960s, structural measures to modify flooding, such as the construction of levees and dams, dominated efforts to reduce flood damage in the larger watersheds of the United States. Among the earliest efforts was the construction of levees in the Lower Mississippi Basin, first by individual landowners and later by levee boards. Beginning in the early part of this century, greater reliance was placed on the construction of large reservoirs.[1] As of 1910 there were fewer than 300 large reservoirs in the United States with a combined capacity of 14 million acre-feet (MAF). In 1988 there were over 2,700 reservoirs in that size class, and their combined normal storage was 467 MAF (Ruddy and Hitt, 1990), though it is acknowledged that flood control is only one purpose served by most of these multiple-purpose facilities. Major growth in these facilities began in the 1920s and peaked about 1965, as shown in Figure 2.2.

---

[1]Large reservoirs are defined by USGS as those with a nonflood control capacity of at least 5,000 acre-feet or a total capacity of 25,000 acre-feet or more.

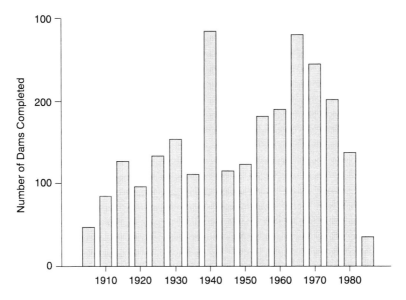

FIGURE 2.2 Number of large dams completed in successive five-year periods in the United States. Source: Ruddy and Hitt (1990).

The sharp decline in construction after 1965 has been attributed to a number of factors. Water resources professionals and environmental groups began to question the benefits of large dam projects. Pasèóge of the Environmental Policy Act of 1970 gave opponents of such projects an entree to the courts while an increasingly urban U.S. population waned in its support for rural construction projects. At the same time, the costs of dams and reservoirs increased as the most attractive sites had already been developed, and debates over how the costs of building these facilities would be shared between federal and nonfederal partners put many projects on hold.

Nonstructural Measures

In the 1960s, behavioral research by Gilbert White and others began to show that, as the frequency of flood events was reduced by building control structures, protected lands became more attractive for urban development. When the frequent floods did occur, property damage was considerably higher, resulting in even higher average damages.

After a series of flood disasters in the early 1960s, White chaired the Bureau of the Budget's Task Force on Federal Flood Control Policy whose report, *A Unified National Program for Managing Flood Losses*, recommended a broader perspective on flood control that embraced the management of floodplains. Published as House Document 465, that report recommended, among other actions, that:

- flood-prone areas be delineated by an appropriate federal agency,
- uniform techniques be established for defining flood frequencies,
- flood forecasting techniques be improved, and
- the feasibility of a national flood insurance program be determined.

The insurance program was to provide financial incentives for private insurers who would provide protection in flood-prone areas that they would not insure otherwise. It would serve two objectives: first to have property owners bear a share of the risk and respond accordingly and, second, to provide disaster relief through the private sector when damage occurred. When the feasibility study was completed by the U.S. Department of Housing and Urban Development in 1965, its recommendations were translated into policy in 1968 when Congress passed the National Flood Insurance Act (NFIA). Modifications to the law were necessary in 1969 and 1973 to overcome local government hesitancies to adopt sufficiently strict land-use regulations. The Flood Disaster Protection Act of 1973 included a provision that made adoption of those controls a prerequisite for federal financial assistance.

FEMA now administers the National Flood Insurance Program and is instrumental in promoting mitigation of the effects of a wide range of hazards. Under the Disaster Relief Act of 1974 as amended and the Stafford Disaster Relief and Emergency Assistance Act of 1988, FEMA is authorized to provide financial assistance to individuals, state and local governments, certain nonprofit organizations, and Indian tribes. The Hazard Mitigation Grant Program, triggered by a presidentially declared emergency, is a federal cost-sharing program authorized under the 1988 act that builds hazard mitigation into postdisaster recovery operations. A major limitation of the National Flood Insurance Program has been its exemption from flood insurance purchase requirements of those structures built prior to enactment of the program. FEMA is also authorized to purchase flood-damaged property and to offer owners an opportunity to relocate.

A number of agricultural policies also are directed at hydrologic hazards. Of special importance is the Federal Crop Insurance Program first established under the Agricultural Adjustment Act of 1938, a program that offers farmers flood insurance for at-risk production areas. The U.S. Department of Agriculture offers a variety of other financial and technical assistance programs. Some, like disaster payments under the Agricultural Consumer Protection Act of 1973 and cost sharing under the Emergency Conservation Program in the Agricultural Credit Act of 1978, are targeted specifically to disasters. Others, such as the Wetlands Reserve Program in the 1990 Farm Bill, "swampbuster" provisions of the Farm Bill of 1985, and the small watershed program under the Natural Resources Conservation Program, are directed at modification of flood events through soil conservation practices, structural measures, and protection of strategically located wetlands.

The infrastructure of dams, levees, and floodways built by the USACE is well known, but the Corps has many other flood management activities. Its Flood Emergency Operations and Disaster Assistance programs provide a wide variety of flood-fighting and rescue operations, assistance for repairing flood control works, emergency water supplies, and other services. Its Floodplain Management Services Unit provides nonemergency technical assistance, including flood hazard mapping and planning.

The Economic Development Administration and the Small Business Administration are significant in the provision of financial assistance to economic development activities and businesses adversely affected by natural disasters. They have only modest roles in modifying the hazard or the risk of damage.

FIFMTF conducted an assessment of progress under NFIA in 1992, noting a long list of achievements—more widespread public perception of the hazard; improved knowledge, standards, and technology; an extensive body of judicial decisions; and well-established development standards. Measuring program effectiveness remained an elusive task, however. The report pointed to the lack of " consistent, reliable data about program activities and their impacts" as a principal complication (FIFMTF, 1992, pp. 60-61). It noted that susceptibility to flooding in the United States is being reduced at individual sites and local communities through a variety of land-use controls and emergency preparedness activities. Evidence reviewed by the task force led its members to conclude, however, that overall vulnerability has either increased or remained the same because of the large amount of preexisting vulnerable development, numerous exceptions in state and local policies, or the inability of governments at all levels to respond quickly to new spurts in development activity.

Following the disastrous floods of 1993 in the Mississippi-Missouri River basin, the Interagency Floodplain Management Review Committee (IFMRC) was established in January 1994 to investigate the causes and consequences of that flood, to evaluate the performance of existing floodplain and watershed management programs, and to make recommendations for appropriate changes. Among the committee's numerous findings was that initial estimates of those properties actually covered by flood insurance ranged from below 10 percent up to 20 percent of insurable buildings in identified flood hazard areas in the Midwest. For the nation as a whole, the range is 20 to 30 percent. The committee recommended taking more vigorous steps to market the program and to provide reduced postdisaster support for those who are eligible but do not purchase adequate coverage (IFMRC, 1994, pp. 131-134).

DROUGHTS

Most extreme events of nature that result in large economic losses or large numbers of deaths are relatively sudden events of short duration, for example floods, hurricanes, and earthquakes, but droughts may have equal or greater consequences. Events that signal the beginning of drought disasters are more gradual than more violent disasters, but droughts may persist for several years, as did the most recent one in California, which lasted from 1987 through 1992. For a variety of reasons (discussed later in this section), damages from droughts are also much more difficult to measure than those of other extreme events that occur within short periods of time.

A fundamental definition of drought does not exist. There is no drought analysis method that utilizes a probabilistic method that is as well developed as flood frequency analysis. While low-flow analysis is a meaningful measure of drought, it does not provide a sufficient definition of drought. Four different working definitions of drought have been established: meteorological, hydrologic, agricultural, and economic drought. All four are based on the notion of water deficit but differ depending on the perspective of the water user.

Meteorological drought can be expressed solely by the degree of dryness and the duration of the dry period (Wilhite, 1993). Normally, the precipitation deficit over a specified length of time determines the degree of dryness. Meteorological drought is very region specific because the atmospheric conditions that result in precipitation deficiencies vary greatly from region to region (Wilhite, 1993).

Hydrologic drought is associated with the effects of precipitation deficits on streamflows, reservoir and lake levels, and ground water supplies. That is, the impact of this deficiency is assessed throughout a hydrologic system. Hydrologic droughts are usually out of phase or lag the occurrence of meteorological drought. Hydrologic droughts may impact local water supplies, hydroelectric power production, flood control, irrigation, commercial navigation, and recreation.

Land use plays an important role in the determination of hydrologic drought. Changes in land use not only significantly alter the hydrologic characteristics of the immediate watershed but can also greatly impact downstream areas. Increased variability in streamflow owing to urbanization effects will cause an increased likelihood of hydrologic drought downstream. Another factor that influences hydrologic drought is the water supply for human and industrial consumption: periods of droughts cause an increased water need, further depleting water supplies (Wilhite, 1993; Grigg, 1996).

Agricultural drought occurs when soil water is inadequate to initiate and sustain normal crop growth over a substantial period of time. Although agricultural drought focuses primarily on soil water deficits, it also incorporates precipitation shortages, differences between actual and potential evapotranspiration, and reduced ground water or reservoir levels.

Economic drought relates to the supply and demand of some water-related good or service and concerns the areas of human activity affected by meteorological, hydrologic, or agricultural drought. Therefore, economic drought is strongly based on human needs (Changnon, 1989). The incidence of economic drought could increase because of a change in the frequency of physical events, a change in societal vulnerability to water shortages, or both.

Because of difficulties of measurement, historical information on the frequency, magnitude, and duration of droughts is far more limited than that of floods and other hazards. Much of the information is anecdotal—the Dust Bowl days of the 1930s, droughts of Southern California in 1927 to 1932 and 1987 to 1992, the drought in the Northeast in the 1960s, and the nationwide drought of 1988. Much of the economic information is also specific to particular effects on agriculture, public water supplies, hydroelectric power, and navigation. Very little information has been compiled to estimate average annual losses across all sectors of the economy.

One of the more complete descriptions of drought frequency at the national scale was produced by Riebsame et al. (1991). They calculated a 90-year sequence of area-weighted annual precipitation from 1895 through 1985 (see Figure 2.3) and similar series for selected seasons. For each year of that same period of record, they also calculated the percent of land area in extreme drought as indicated by the Palmer Hydrological Drought Index (see Figure 2.4) and accumulated precipitation deficiencies. All of these measures were calculated for the contiguous United States. The same measures could also be calculated for selected regions of the country.

When droughts occur, their impacts are quite diverse. As part of the National Drought Study (USACE, 1994), the Corps of Engineers surveyed all 50

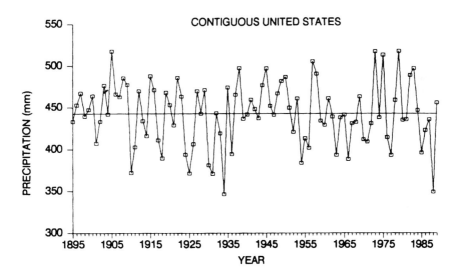

FIGURE 2.3 Total annual precipitation, 1895 to 1989, area weighted over the contiguous United States. Source: Riebsame et al. (1991).

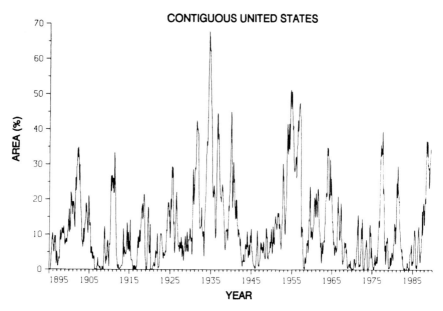

FIGURE 2.4 Percentage of the contiguous United States in severe or extreme drought (Palmer Hydrological Drought Index ≤3), 1895 to 1989. Source: Riebsame et al. (1991).

state governors to identify the types of impacts that affected their regions. Shortages of public water supplies and crop losses were among the most frequently cited effects. Environmental damages to wildlife, fisheries populations, and other aquatic ecosystems were also frequently mentioned. Impacts on the Everglades were of special concern. In the Northwest, loss of hydropower was especially important, and in the Midwest, interruptions to navigation on the Ohio and Mississippi rivers were cited. A number of states cited effects on aquifers that were subject to saltwater intrusion. Economic losses for the 1988 drought as estimated by the Interagency Drought Policy Committee totaled $39.2 billion, over 60 percent of which was attributed to losses in farm production and increased food costs (Riebsame et al., 1991).

The National Drought Study identified a number of difficulties in measuring the consequences of droughts. Agricultural losses in one region of the country can be offset by increases in other regions. Regional industrial losses can at least be partially offset by increases elsewhere. Because droughts extend over relatively long periods of time, separating their effects from others such as shifts in the general economy and changes in management practices is often difficult. In addition, techniques that focus only on losses during droughts fail to account for capital expenditures made in earlier years to mitigate the effects of droughts, including the building of reservoirs for public water supply, hydroelectric power, low-flow augmentation, and navigation.

DROUGHT MANAGEMENT STRATEGIES

Unlike that for floods, no unified national management strategy has been developed for droughts. By contrast with specific strategies outlined by the FIFMTF, the *National Study of Water Management During Drought* (USACE, 1994) contains only a broad planning framework for drought management.

That is not to say that droughts have been ignored. Substantial capital investments have been made by USACE, the Bureau of Reclamation, the Tennessee Valley Authority, the Natural Resources Conservation Service, and other federal agencies to develop conservation storage for hydroelectric power, navigation, municipal and industrial water supplies, and low-flow augmentation. Furthermore, there has been a substantial history of federally legislated financial assistance to farmers for drought relief both in general and in particular years.

Local governments and the private sector have likewise made substantial investments in reservoirs to augment flows for public water supplies and hydroelectric power. About 85 percent of all municipal water supplies in the United States are provided by local governments, and a substantial share of that is taken from surface water reservoirs. In some cases those supplies are provided by federal reservoirs, and in California the state has made a large investment to develop water supply reservoirs and distribution systems.

A special challenge facing all operators of surface water and ground water reservoirs is how to allocate available resources during a drought. While these facilities are usually designed to provide a given yield under drought conditions of specified frequency as determined from historical records, operators face considerable uncertainty during any drought as to what its magnitude and duration will be. They are usually confronted with the need to hedge against the possibility that a particular drought may be more severe than that for which the system was designed.

USGS ROLE IN HYDROLOGIC HAZARDS

The USGS provides several types of support to water resource managers and emergency management officials for addressing the issues associated with extreme floods and droughts, including (1) determining the probability of occurrence of extreme hydrologic events; (2) improving our understanding of the processes that determine the severity of extreme events, including anthropogenic factors; (3) understanding ancillary impacts of extreme hydrologic events such as bridge scour associated with floods; (4) providing tools for assessing alternative strategies for mitigating the impacts of floods and droughts; (5) monitoring of ground water levels and streamflow conditions; and (6) undertaking assessments of the magnitude and extent of floods and droughts in support of emergency management officials.

The long-term streamflow and ground water-level monitoring programs of the USGS provide the base information for determining the probability of occurrence of extreme hydrologic events. The current network of 7,000 daily streamflow stations and the more than 27,000 other stations that have previously been operated as daily discharge stations or peak discharge or low-flow stations provide a robust database for assessing flood and drought potential in many parts of the nation. It is not practical or economical to monitor every stream, so the USGS uses statistical techniques to estimate the probability of occurrence of floods and low streamflow of various durations for ungaged streams. These techniques typically employ regression equations for estimating specific flow characteristics based on physical characteristics, such as area of the watershed, average annual precipitation, slope of the stream channels, land use, and amount of water storage (lakes and reservoirs) in the watershed. The USGS's National Flood Frequency Program should soon be available on the World Wide Web for estimating peak flow characteristics throughout the nation. Regression equations for estimating low-flow characteristics are available in reports published by individual USGS district offices, generally in cooperation with the primary water management or natural resource agency of each state.

A subelement of the probability of occurrence of floods and droughts is understanding the processes that determine the severity of a flood or drought. The climate conditions prevailing at the time of the event and antecedent conditions are major factors affecting the magnitude or severity of a flood or drought, but sometimes even these factors are obscured by human activities such as land-use practices, long-term water use, and changes in management of river systems. The USGS periodically conducts assessments of how streamflow characteristics change because of anthropogenic or climatic conditions. For example, many of the states that have regression equations for estimating flood characteristics of streams with undisturbed or natural watersheds also have ancillary equations for estimating flow characteristics of streams with watersheds with varying amounts of urban development. The USGS also has conducted studies to determine the effects of deforestation, drainage for improving crop productivity, and other land-use alterations.

Impacts of flooding besides inundation include deposition of sediments in river channels, reservoirs, and floodplains and scour of river channels, particularly the foundations of bridge piers and abutments. The USGS monitors streams for sediment transport and has conducted various studies to determine the variation of sediment yield from watersheds with different land-use characteristics. Determination of the potential for large-scale input of sediment to streams from landslides and channel erosion is addressed as part of this activity. This information is used for identifying areas subject to extensive sediment transport and deposition to optimize the location of water storage and treatment facilities. It

Discharge measurements are especially important during extreme flow events, even though conditions are least favorable. Photo courtesy of U.S. Geological Survey.

also provides a basis for designing flood control facilities, including providing adequate storage for water control and sediment accumulation. The USGS also has been conducting research on stream and bridge site characteristics that affect the potential for scour of the foundations of bridge piers and abutments. Physical characteristics such as channel slope and size of streambed materials have been used in conjunction with scour measurements to develop equations for estimating the potential for scour at existing or planned highway bridges. The research also has resulted in guidelines for state highway departments to consider while inspecting and designing bridges.

The development of tools to help resource managers and emergency officials plan efficient mitigation strategies is another aspect of USGS activities related to hydrologic hazards. The agency has developed various numerical models and other analytical techniques for predicting runoff from watersheds and for routing flood peaks from the headwater streams to major rivers. Watershed models are being used for assessing various alternatives for operating reservoirs for reducing flood hazards and strategies for distributing water to water-deficient areas during droughts. Watershed models also are used for evaluating nonstructural alternatives for reducing the impacts of floods. The WSPRO (Water-Surface Profiles) model developed in the mid-1980s has been used extensively to determine areas subject to inundation for floodplain management and flood insurance rate determinations. WSPRO and multidimensional models such as FESWMS (finite-element surface-water modeling system) also have been used to design culverts and bridges at single and multiple stream crossings of local, state, and interstate highways. Improved understanding of fluvial geomorphology through USGS research has resulted in sediment transport models for assessing the effects of floods and water regime modifications on aquatic and riparian ecosystems.

The USGS role in collecting and distributing real-time streamflow and ground water-level data has expanded dramatically in the past two decades. There have been long-term agreements with the National Weather Service, the Army Corps of Engineers, and other agencies needing data for forecasting or operational decisions to provide access to the data through telephone or radio links to USGS monitoring sites. In the 1980s the USGS's capability to distribute hydrologic data directly to resource management and emergency management organizations was enhanced with the availability of instruments and satellites for transmitting data from remote locations. Today, the USGS serves real-time streamflow data from more than 3,000 monitoring sites to the World Wide Web via Geostationary Operational Environmental Satellites (GOES). These data are now available to resource managers, local emergency officials, and individual property owners who need to make decisions on how best to deal with either a flood or a scarcity of water.

The other area where the USGS plays a significant role in hazard programs of various organizations is in rapid assessment of the magnitude and extent of floods. USGS personnel respond to extreme hydrologic hazards by collecting critical data during and immediately after the events to characterize the extent of flooding. This information includes the determination of flood stages and discharges to classify the floods for disaster assistance and insurance purposes. It also includes documentation of flood profiles and flood inundation maps for both short- and long-term planning. Emergency management officials use the information to make informed logistical decisions for distributing flood-fighting resources, rescue teams, potable water, food, medical supplies, and temporary housing. The information is also used for land-use planning and for assessing floodplain management strategies.

ISSUES AND IMPLICATIONS FOR INFORMATION NEEDS

A number of issues arise out of the national strategy to manage floods and droughts, and the need for a continued flow of information to support the strategy is readily apparent. This report addresses issues directly related to the USGS. Many of those issues and information needs are not new but are worthy of restating nonetheless.

For the management of floods, there is a need for continued improvements in methods for estimating the frequency and severity of extreme events and their consequences. In addition to a need to continue efforts to improve techniques for long-range forecasting of floods, at least two other avenues are worthy of pursuit. First, on regulated streams and in urban areas and other places where hydrology is being modified by land development, the probabilities of flood peaks and volumes are being altered, often dramatically, as demonstrated by a number of field studies. The USGS periodically assesses those impacts using regression equations, but the generalization of results needs improvement. Hydrologists need enhanced procedures for adjusting the probabilities of extreme events in a timely manner. A second issue is the lack of readily available and up-to-date information about the consequences of extreme events for which probabilities have been increased by development activity. Despite technical information and warnings, communities are often caught off guard by flood events. The reasons for this are unclear but there are several contributing factors. Maintaining "readiness" requires a commitment of financial and other resources. Perhaps also local officials do not understand probabilities very well or the consequences of an event of given probability; perhaps the information is communicated in technical language not readily understood by the affected communities. Unfortunately there are also many instances where risk is well understood but ignored by local officials who may act unwisely (as, for example, local planning boards are often subject to political pressure to develop floodplain lands).

Flood management also requires conditional forecasts of flood events given an occurrence of exceptional precursor events. Many flood events, like those in Northern California and the Red River, follow periods of exceptional snowfall. Others follow periods of sustained rainfall that leave soil in near-saturated conditions. Knowledge that events of this kind have occurred can substantially increase the probability of flooding in comparison to long-term probabilities. Such information is of significant value to flood preparedness activities.

In the best of times, budget resources for the collection of basic information are limited, and the stream-gaging program is no exception to that rule. In an era of downsized government, every noncritical expenditure has come under closer scrutiny. Although the gaging program is critical, especially during extreme events, justification for continued support of the program has become more difficult in recent years. Such pressures continue to underscore the need for a more economically efficient means for improved data collection. Recent technological advances have led to improved methods for gathering, storing, and transmitting data, but there is little evidence to suggest that the cost of data acquisition has decreased. Improved techniques for regionalizing the results of available gages also could be enhanced.

The hydrology of droughts is much less well understood than the hydrology of floods. Unlike flood hazards, where flood-prone areas have been delineated to some degree at least by the probability of inundation, there is no systematic process for estimating the probabilities of drought. Droughts are defined differently for different purposes. An agricultural drought is not necessarily the same as a municipal water supply drought. It may be necessary to establish different definitions of droughts and different estimates of probabilities to respond to different needs. Improved techniques for estimating drought probabilities are needed, as are improved methods for communicating those probabilities and related consequences to the public.

As with flooding, conditional forecasting of droughts could be helpful. It is one thing to know the long-term probability that a drought will occur in a particular region. It is another to know the probability of the duration and magnitude of a drought once it is recognized that one has started. Such information is crucial to the allocation of resources during drought events, to decisions whether to continue investments in crop production, or to decisions about the movement of goods by navigation or alternative means. Incorporation of knowledge about prevailing large-scale meteorological conditions and probabilities of when those conditions will change could produce more useful statements of probability.

3

Data Collection,
Techniques Development, and Research

The U.S. Geological Survey (USGS) performs critical functions in relation to hydrologic hazards and extreme hydrologic events (see Figure 3.1). These functions encompass broad areas of data collection, techniques development, and research. Routinely, or "prior" to a major event, such as a flood or drought, the USGS collects and manages baseline data; constructs predictive models; develops new modeling, data analysis, and instrumentation methods; and helps communicate hazard information to user groups through maps, reports, public contacts, and other means. During an extreme event the USGS has a critical role in real-time data collection, data management, and communication of hazard information. After an event, the USGS documents what happened with the goal of understanding why the event occurred and identifying problems in modeling, data collection, and understanding of the physical processes. The information and experience gained through these functions should enhance and improve the understanding of extreme events.

Research, technique, and data needs relative to hydrologic hazards fall under two main categories—floods and droughts. Aquatic ecosystems rely on water, in the appropriate balance, for their well-being. Hydrologic hazards such as flooding and drought can upset the balance needed for some aquatic ecosystem while providing necessary periodic events to maintain other systems. Analytical difficulties that encompass most of hydrology and require added research are problems associated with the stationarity of hydrologic processes, the spatial and temporal variability of hydrologic processes, and issues related to the scaling of hydrologic processes. These three universal needs are discussed in detail in other reports of the National Research Council (NRC) (1988, 1991) and remain valid

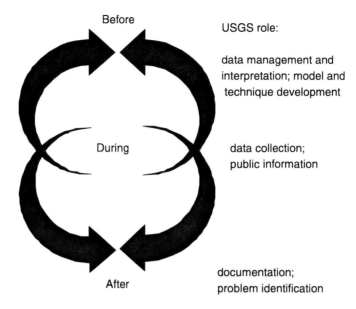

FIGURE 3.1 Roles of the USGS in extreme hydrologic events.

today. The collection of reliable, relevant datasets will continue to be a critical component of the hydrologic hazards program. Finally, there is a vital need to find better ways of communicating the consequences of hydrologic hazards to the public. The USGS is poised to take the lead in the development and application of improved visualization techniques that exploit the agency's expertise in long-term monitoring, mapping, and process modeling.

FLOODS

Floods are the most visible and immediately destructive extreme hydrologic events. Floods create hazards in several ways:

- loss of life, property, and services and other social impacts associated with flood flows;
- scour and deposition in natural channels, including scour around bridge piers and other structures located in the floodway;
- erosion of channel banks and beds and deposition of sediments in channels, impoundments, and other areas;
- possible transport of large quantities of pollutants;
- landslides and debris flows; and
- alteration of hydrologic characteristics of basins.

Flood overtopping Sorlie Bridge, Grand Forks, ND, during spring of 1997. Photo courtesy of U.S. Geological Survey.

In addition to their human impacts, floods can have severe effects on natural ecosystems. During floods, destruction of natural habitat occurs through channel erosion and excessive sediment deposition, typically exacerbated by increased peak flood flows associated with urbanization and by increased erosion associated with some agricultural practices, water management, construction activities, and other earth-moving operations.

Adverse water quality conditions can result from stormwater runoff, both within and downstream of urban, agricultural, silvicultural, and mining areas. Constituents of concern include high sediment loads, metals, nutrients, pesticides and herbicides, and other organic contaminants.

Assessing the potential impacts of floods requires good estimates of the magnitude and frequency of flood flows. A 1988 report of the NRC, *Estimating Probabilities of Extreme Floods: Methods and Recommended Research,* is devoted entirely to this issue. The recommendations set forth in the concluding chapter of that report remain valid. The report indicates that the theoretical basis for estimating flood probabilities needs improvement, possibly through the application of regional statistical analyses of streamflow, consideration of paleohydrologic data, development of a comprehensive statistical model, and the use of rainfall-runoff models driven by synthetically derived precipitation inputs. Advances have been made in all of these areas since the 1988 report, but much remains to be done and the recommendations remain valid.

Regional estimates of flood magnitudes and frequencies have become standard tools for flood analysis. Political boundaries often influence the definition of "regions." Such limitations are artificial and can be overcome through management of already available data. The concept of defining regions based on geographic location has advantages in terms of selecting the regional hydrologic relationship to use in a particular analysis but may not be optimal in terms of reducing the standard error of the estimates of flood peaks. The definition of regions based on hydrologically related similarity criteria may offer improvements in terms of reduced standard errors (Tasker et al., 1996). Such regional definitions related to the hydrologic responses of watersheds should be continued and expanded to include hydrologic characteristics such as climatic variables and soil and geologic variables in addition to the commonly used physical characteristics of watersheds.

Human activity has altered the magnitude and frequency of rare events. Constructing dams and levees, altering drainage characteristics, and changing land use all can introduce nonstationarity into hydrologic records. Currently, we do not know the potential for an extreme hydrologic event to override human control of rivers and/or channels and thus result in a disaster of a magnitude far beyond what would have occurred under less managed conditions.

Data Needs Related to Floods

Data and network design questions, with an emphasis on real-time data collection, are key issues in understanding, predicting, and documenting floods. All predictions and models must be grounded in appropriate and reliable field data, and the USGS should give data collection the highest priority in hydrologic hazards programs. In particular, the agency should collect data that capture the variability of high-flow hydrologic events in space and time. An adequate national stream-gaging network must provide data for hydrologic warning in real time, for analysis of historic events, and for documentation of rare extreme events. Detailed studies of extreme flood events are essential for predicting future events of a similar nature. An adequate database to support such studies requires sufficient site specific information to elucidate the critical hydrologic, hydraulic, geomorphic, and hydroclimatic factors that shaped each extreme event. Data needs, therefore, transcend traditional gaging (and agency) activities. The greatest value will result from the synthesis of land use, hydraulic modification, river regulation, meteorology and climate conditions, with gage data and flow measurements. Such a synthesis may be readily achievable by coordinating the activities of the Corps of Engineers, NOAA, USDA, and USGS through a refocused concentration on hydrologic hazards.

Improved understanding of extreme events may require new approaches to collection of streamflow data. The traditional USGS paradigm for stream gaging has been to collect highly accurate information at a few distinct geographic sites.

One weakness of the present streamflow monitoring system is a lack of stream-flow and water-level data in the areas of highest hydrologic variability, such as steep mountain basins and arid arroyos of the Southwest. Also, more emphasis should be given to obtaining data in urban areas, and data to better understand the impacts of water project operations would be instructive. Another current weakness is the program's emphasis on measuring of average flows rather than the rare extreme event. This is a daunting problem because it will be difficult for the USGS to justify data collection efforts (e.g., the installation of gaging equipment far above the usual high-water stage) that are used only rarely, yet collection of data during extreme events is critical to improved understanding, prediction, and warnings. It may be appropriate for the USGS to evaluate the role of peak stage recorders as part of an alternative data collection paradigm of collecting slightly less accurate information at more geographic sites.

Networks designed for characterizing regional hydrology may not be optimal for providing real-time data for hazard warning systems such as flood warnings. Added rain gages with the ability to transmit data in real time to a receiving location may be needed as real-time flood warning systems begin to rely on radar rainfall estimates. A growing infrastructure of real-time streamflow and precipitation data collected by state and municipal agencies may valuably augment the Survey's core data collection programs. Interagency cooperation to establish common formatting and quality control standards may be a cost effective means to expand the utility of regional hydrometeorological databases for diverse applications such as ground truthing and calibration of weather radars. The raingages are required to provide data that can be used to calibrate the radar echoes.

Flood research and forecasting requires both streamflow and water quality data for developing and testing hydrologic analysis techniques and for estimating parameters required in hydrologic and water quality models. Major runoff events can introduce large quantities of pollutants into surface waters. Because of the cost of collecting and analyzing water quality samples, there are far fewer locations collecting time series of water quality data than there are for streamflow data. Improving the database of water quality data will require less costly methods of data collection and analysis.

Technique Development Related to Floods

Many opportunities exist for development of techniques related to the measurement, forecasting, and simulation of floods. The most basic of these is the actual measurement of flood discharge and water stage. At high-flow rates, channel cross-sections, especially the bottoms of natural channels, can become unstable, resulting in scour and/or deposition of sediments that alter the relationship between flow rate and flow depth. Channel instability is especially acute in arid and semiarid regions. Extreme events are also difficult to measure because of the adverse conditions that exist and the hazards presented to personnel making the measurements. Once water levels overtop normal banks, it may no longer be feasible to actually measure the flow. For example, bridges are often used as platforms from which actual streamflow measurements are made. Extreme flows may overtop the roadway on either side of a bridge or even on the bridge itself, making the bridge inaccessible and inadequate for making measurements over the total width of the flow. Improved methods of streamflow measurement are needed that are less labor intensive and can be carried out quickly without the need to repeatedly physically lower instruments on a cable into the water. Instruments are difficult to position in high flows and are subject to entanglement in debris. Semiarid and arid regions present other challenges in that flow rates may go from near zero to record-high levels in a matter of minutes. Current manual streamflow measuring techniques are not suited to making measurements under these conditions.

Techniques should also be improved for the construction of stream rating curves, which are often quite uncertain for extreme events. Generally, such rating curves are extrapolations of rating curves developed from more normal flows.

As greater dependence is placed on real-time data, reliability and redundancy will need to be improved to ensure that hazard warning systems remain operational at critical times. If a key stream gage becomes inoperative during a major flood event, a flood warning system could be rendered ineffective, resulting in more damages and losses than might have occurred using older technologies.

Research Needs Related to Floods

Data on relatively frequent peak flows are naturally far more readily available than data on extreme events, which by definition occur infrequently and likely are not represented in commonly available records of length. Statistically derived flood flow estimates assume stationarity and homogeneity, meaning that extreme event are from the same statistical population as the more frequent events and thus their magnitude and frequencies are imprinted in these more frequent events. However, flood flows are probably not statistically stationary (Klemes, 1986). Can the magnitude and frequency of extreme events be predicted based on a statistical analysis of frequent but relatively small flood flows thought to have return periods of a few years? Hydrologists seldom have adequate data for a statistical treatment of floods in the more normal range of return periods (<100 years). There are few, if any, records that can be used to directly estimate 1,000-year events via standard flood frequency analysis.

How do mixed-population statistical techniques apply to flood frequency analysis? Traditional flood frequency analysis relies on the assumption that all flood peaks are from the same population. Yet we know that in many areas flood events may arise from two or more hydrometeorological conditions. For example, in the southeastern United States, flood flows might arise as the result of frontal systems or hurricane systems. The magnitudes and frequencies of flows from these two sources may be quite different, indicating that combining them into a single peak flow sequence may be inappropriate.

Continued research is needed to develop and improve techniques for extrapolation of hydrologic data and predictions to extreme events. The use of stochastically generated rainfall events implicitly relies on the same assumption inherent in statistical flood frequency analysis—that extreme rainfalls follow the same statistical patterns as the more frequent rainfalls that are the basis for estimating the parameters of stochastic models. Routing statistically generated rainfall through a rainfall-runoff model to estimate extreme hydrologic events also contains the assumption that the hydrologic response to extreme rainfalls can be modeled in the same way as the response to the more frequent rainfalls used to calibrate the model.

Improved modeling may be one solution to better understanding and prediction of flood events and will requires a mix of real-time dynamic modeling and retrospective deterministic and stochastic modeling. Is there any pattern to hydrologic events such that events preceding a hazardous event can be recognized and thus aid in the prediction of a major hydrologic event? Can we identify signatures in hydrologic signals that when combined in certain ways lead to potentially hazardous situations? This would require detailed analyses of past hazardous events. It also requires data adequate to characterize an actual event and the conditions leading up to it.

Research is also needed to improve network design. What are the data needs for real-time flood forecasting? How can gaging programs be optimally designed to provide these needed data? Possibly the examination of geologic records can be used to determine the magnitude and frequency of rare events. We know there have been dramatic climatic shifts over geologic time. How did these shifts impact hydrology? How can past climatic changes be factored into extreme flood event estimation using paleofloods?

Understanding the causes and consequences of floods requires continued research on the geomorphology of stream systems. Relationships between bank stability and bank erosion, depositional patterns, scour patterns, and meander characteristics need to be studied as a function of stream geology and hydrology and hydraulics and how changes in these factors induce changes in the stream. Such problems are particularly severe in urban areas, where formerly stable channels become unstable as a result of increased flood peaks and/or flow volumes and reduced sediment loads. A wealth of information on geomorphology could be extracted from the USGS's vast discharge measurement file.

It is often unclear how to evaluate the significance of impacts from stormwater runoff on aquatic ecosystems, including environmental indicators as well as traditional water quality parameters. For instance, the Center for Watershed Protection (a nonprofit organization in Maryland) recently released a proposed methodology for applying environmental indicators to the assessment of stormwater quality control programs, because chemical parameters are insufficient. Twenty-six indicators are proposed, within several categories, including water quality (toxicity testing and chemical conditions); physical and hydrologic (e.g., physical habitat, flooding frequency, and dry weather flows); biological (e.g., fish and macroinvertebrate assessments); social (e.g., public surveys); programmatic (e.g., Best Management Practices installed); and site-specific (e.g., BMP performance). The potential for the Biological Resources Division of the USGS to get involved in the evaluation of stormwater impacts could be considered.

DROUGHTS

Droughts can be thought of as creeping disasters, and often the public is not convinced that a drought exists until it reaches its most severe stages (Matalas, 1991; Grigg, 1993, 1996). The nation's 1988 drought cost an estimated $40 billion in damages and 5,000 to 10,000 deaths (includes heat-related deaths) (NDMC, 1997). The losses resulted from reduced crop yields, severely disrupted barge traffic, and critically low municipal water supplies (Kunkel and Angel, 1989). Environmental impacts of drought also are of great concern. Plants with shallow root systems are damaged or killed and at best have zero growth. There were also major reductions during 1988 in wildlife populations, such as pheasant, ducks, and fish. Because of the lack of surface runoff and low in-stream flows, nutrient fluxes ceased in the riparian corridors. Lack of nutrients and above-normal temperatures combined to cause high mortality of aquatic life. On a global scale, the occurrence of droughts historically has had great impact on the political stability of affected regions, as drought has led to food shortages, starvation, and the mass migration of peoples.

Dracup et al. (1980) proposed the following set of criteria to be considered when defining drought: (1) Is the primary interest in precipitation (meteorological drought), streamflow (hydrologic drought), or soil moisture (agricultural drought)? (2) What is the fundamental averaging period of the time series to be studied (e.g., months, seasons, or years)? (3) How are drought events distinguished analytically from other events in the time series? (4) How are the regional aspects of droughts to be considered? Once the nature of the water deficit has been selected, the time or duration to be used as an averaging period must be chosen for the meteorological and hydrologic variables under consideration. For the study of drought, this can range from months to seasons and years. The period chosen will directly impact the data's sample sizes, which will impact the number and type of drought events. For a given hydrologic record, a shorter averaging period results in a larger number of drought events, while a longer averaging period results in a smaller number of drought events. In this context the regionalization of drought is extremely important. Droughts are inherently regional in nature, and thus their areal extent is an important characteristic to be considered (Karl, 1983). Another important factor is that due to the small sample size of drought events, drought analysis may be hampered; regionalization on the other hand, provides a means for increasing the sample size.

Drought is also strongly related to the climatological and geological characteristics of a region, and a general drought definition can be considered to be a water shortage with reference to a specified need for water in a conceptual supply and demand relationship (Dracup et al., 1980). The occurrence and persistence of droughts is linked ubiquitously to high surface temperature anomalies. Over the last century, the driest periods in the central United States have consistently occurred during some of the warmest periods of the century. The 1988 drought is a prime example: it ranked sixth in dryness and first in warmth for the central region of the United States over the past 100 years (McNab, 1989).

Atmospheric circulation also plays a major role in drought severity. The occurrence or absence of precipitation on any individual day can be related to daily atmospheric circulation patterns. Days with little or no precipitation during both drought and nondrought periods can be attributed to the absence of destabilizing temperature or velocity gradients or to widespread descending vertical motion (McNab, 1989). These precipitation-inhibiting circulations recur more frequently or persist longer during drought periods than nondrought periods. It is the persistence or frequently recurring character of these circulation patterns that is unique to droughts (Namias, 1985). Prolonged droughts occur when large-scale anomalies, such as those associated with ENSO, in atmospheric circulation patterns persist for months, seasons, or years (NDMC, 1995, 1997). Hydroclimatological features, which describe the spatial and temporal connections between atmospheric circulations and the hydrological cycle at the regional scale, can be used to assess the predictability of drought severity and duration. Regional anomalies of precipitation, temperature and evapotranspiration, and their time evolution are prime candidates for such studies. One complicating factor is natural climate variability and the lack of sufficiently long datasets to be able to separate stationary, nonstationary, and random components in hydrological time series.

The amount and duration of the deficit influence the severity of the drought (Changnon, 1989). In 1965 Palmer developed an index to measure the departure of the moisture supply. To date, the Palmer Drought Severity Index (PDSI) is the most commonly used drought severity indicator (Hrezo et al., 1986). The main objective of the index was to provide a standardized measurement of soil moisture conditions, so that comparisons using the index could be made between different locations and different months. The PDSI is a "meteorological" drought index that responds to weather conditions that have been abnormally wet or dry. An index value of +4.0 indicates an extremely wet period, while an index value of −4.0 indicates an extremely dry period. Values usually fall somewhere in between. The PDSI is determined monthly for the 344 climatic divisions of the United States. Each climatic region is supposed to exhibit locally homogeneous climate and geographic characteristics. The PDSI is a commonly used tool because it provides policymakers with a measure of the abnormality of a region's recent weather patterns. It allows current conditions to be viewed in a historical perspective, in addition to providing spatial and temporal representations of historical droughts. Nonetheless, the widespread application of the PDSI has been the target of much scientific discussion, especially with regard to its transportability from one region to another (Alley, 1984). Some of the index's shortcomings are that (1) it does not apply to the definitions of hydrologic or agricultural drought; (2) the values quantifying the intensity and timing of drought were arbitrarily based on Palmer's study of Iowa and Kansas and may not be adequate for other regions of the country; (3) it is sensitive to the available water content as a function of soil texture, and therefore use of the PDSI for an entire climatic division may not be appropriate; (4) snowfall and snow accumulations are not included in the index; and (5) the natural lag between rainfall and runoff response is not considered (NDMC, 1996). For these reasons the PDSI cannot be used to predict the onset of drought, and it is useful only in determining the stage or severity of a drought episode.

On the other hand, a strong need exists to establish a robust paradigm to evaluate the risk of drought. A probabilistic approach to drought analysis as used in flood frequency analysis does not exist. While low-flow analysis is a meaningful measure of drought, it is not a sufficient definition.

Data Needs Related to Drought

Currently, a scarcity of comprehensive datasets limits scientific understanding of the causes, onset, and effects of droughts. Unlike floods, droughts evolve slowly and may not be recognized until they are almost over and the opportunity for data collection is lost. The USGS should develop coordinated plans for data collection related to droughts in order to provide a comprehensive picture of how a drought develops and evolves. Coordination of such data collection programs with NOAA's climate data collection programs and NASA's large-scale remote sensing activities offers a timely opportunity to enrich the integrated databases supporting research, analysis and management of drought. Key data collected by the USGS that are relevant to drought include low-flow measurements and flow duration measurements in surface water and long-term ground water-level fluctuations. Long-term water-level measurements in lakes and wetlands also are essential to document the effects of drought and to provide data to construct hydrologic models. Monitoring of ground water levels and soil moisture also is extremely important during drought periods, and long-term fluctuations in ground water levels are often an excellent indicator of overall climatic patterns. These measurements need to be combined with climatological measurements of precipitation, temperature, wind, soil moisture, and other parameters in order to create a complete picture of droughts as extreme hydrologic events.

Technique Development Related to Droughts

Drought research and technique development should be undertaken together in order to collect relevant datasets for drought monitoring and prediction. As with high-flow events, understanding droughts may require a shift away from the collection of very high-quality data at few sites to the collection of lower-quality data at more sites. Improved data collection can result from advances in instrumentation for low-flow measurement and soil moisture measurement at remote sites to acquire a regional perspective. Better ability to predict drought periods could be applied to improve water management systems, enabling better planning by water users (e.g., municipalities and agriculture) and better protection of drought-intolerant aquatic systems.

A fundamental aspect of hydrologic drought is the regional character of surface-ground water interactions in different landscapes, especially with regard to lake and stream recharge. Such interactions are natural processes that are strongly affected by regional physiographic, geomorphologic, and climatic attributes, and they typically exhibit slow responses to meteorological forcing. Therefore, the development of objective measures to describe the magnitude and seasonality of the interaction between surface and ground water systems is needed to improve drought forecasting skill on a regional basis for water resources applications.

Research Needs Related to Droughts

Predicting the onset of a drought is the major challenge in drought-related research (Matalas, 1991). Critical science issues underlying this challenge include the need for specific, objective criteria to define the initiation of drought, and the need for quantitative forecasts of drought attributes such as duration, severity, and spatial variability. Since public perception of drought is intimately associated with the specific water needs of different users, it is important to research and develop forecast models that can be used operationally to support existing decision-making procedures.

Since hydrologic drought is normally associated with the effects of precipitation deficits on streamflows, reservoir and lake levels, and ground water supplies, there is a phase lag between hydrologic and meteorological droughts. This improves the potential hydrologic predictability. Research focusing on the linkages between the land and atmospheric processes of the hydrologic cycle should therefore lead to a better understanding of drought phenomena and their predictability.

Research is also needed on appropriate minimal streamflows for ecological health. Water quality-based permit limits for dischargers throughout the United States are based on 7-day, 10-year low flows (7Q10). Many are questioning the appropriateness of this flow statistic to protect aquatic life and our ability to estimate the 7Q10 accurately, particularly for ungaged sites.

4

Interpretive Studies

INTRODUCTION

Interpretive studies performed by the U.S. Geological Survey (USGS) represent the interface between data-gathering efforts and research activities and their application to problems of local, regional, and national significance. Interpretive studies are one of the crucial links between the research and service missions of the USGS through the support and application of research activities in major mission areas. The conduct of problem-specific interpretive studies at the district level offers the opportunity for cooperators and various interested parties to benefit directly from research by the USGS through the application of research results to practical local problems. Local client-driven interpretive studies symmetrically support research efforts by expanding the consistent national database available for research, as well as providing practical testing of new research results in applied management problems. Feedback from these applications, in turn, generates and refines the emerging questions and problems that drive the evolving research program toward timely challenges that are well suited to the USGS's expertise.

The value added through the intersection of USGS interpretive studies and research is well demonstrated in the agency's work in bridge scour and flood peak estimation on ungaged watersheds. The state of the practice in the development of interpretive studies is largely driven by funded cooperators and serves a valuable technology transfer function for the more mature research products. Strategic opportunities to strengthen the role of interpretive studies have been identified in supporting public policy on risk management for hydrologic haz-

Typical damage caused by debris flow. Photo courtesy of U.S. Geological Survey.

ards, planning responsive monitoring networks, and evaluating rapid-onset events. Research needs related to interpretive studies include more rigorous treatment of nonstationarity in hydrologic risk assessment, including nonstationarity and persistence in climatic linkages, changing land use and watershed conditions, and the dynamics of channel morphology and sediment transport. The role of the USGS in interpretive studies includes technology transfer, integrating cooperator needs with the research program, and information generation.

INFORMATION GENERATION

Information on hydrologic hazards is not equivalent to the data used to create the information. Data are measurements made to describe selected aspects of physical, ecological, or socioeconomic systems. For example, measured streamflow data can lead to information on hydrologic hazards but such data by themselves are not hazard information. Instruments, laboratories, and surveys are examples of tools for securing data.

Information on hydrologic hazards is created by the interpretation of data, using particular analytical frameworks and concepts. Data interpretation to create information is done by hydrologists and other scientists, engineers, policy makers, and the general public. These various data users apply differing conceptual frameworks to data as they create information relevant to their particular purposes.

The value of information differs according to the decision-making setting. While statistical decision theory defines information value in terms of the economic gains from improving a particular decision outcome, organizational theorists describe a different value for information. Decision makers always are scanning and monitoring the organization's environment for signals that would require shifts in programs and policies.

To illustrate, tracking national trends in weather-related economic damages might signal a need to make changes to the national flood information program. Such global trends data have little value to the farmer making a planting or harvest decision.

Information on hydrologic hazards may be used to predict and/or understand (1) the probability of hazardous events, (2) the adverse consequences of the events, and/or (3) the complex causal pathways linking events to their hazardous consequences.

FLOOD FREQUENCY ANALYSIS

State of the Practice

Interpretive studies related to flood frequency analysis have traditionally focused on the application of standard hydrologic methods described by the U.S. Interagency Committee on Water Data (1982) to analyze the characteristics and frequency of flooding. Statewide surveys (Clement, 1987; Curtis, 1987; Guimaraes and Bohman, 1991) as well as flood frequency analysis for river basins (Landers and Wilson, 1991) provide resource managers with standard, reliable baseline information on flood frequency that directly supports design and floodplain management activities throughout the nation. Interpretive studies have also integrated the products of the national research program in hydrologic regionalization and the estimation of flood peaks at ungaged sites (Stedinger and Tasker, 1985, 1986; Tasker and Stedinger, 1989, 1992; Tasker and Slade, 1994; Tasker et al., 1996), culminating in national regression equations for flood peak estimation (Jennings et al., 1994).

Flood-related interpretive studies documenting the magnitude and hydrologic characteristics of extreme events expand the national information base on flood risk. These activities utilize hydrologic and hydraulic techniques (Dalrymple and Benson, 1967; Kirby, 1981; Fulford, 1994) as well as paleohydrologic techniques (Costa, 1986; Cohn and Stedinger, 1987).

Nearly all mathematical idealizations of hydrologic processes presume that the underlying probability distribution for a random variable, such as flood recurrence, remains constant through time. Stationarity is an assumption underlying most conventional techniques for estimating flood exceedance probabilities.

However, changes in climate, vegetation, and other watershed conditions are known to drastically alter flood responses in many situations.

For paleoflood records extending back beyond the past century the key stationarity questions center on climate variability. Deviations from purely random event generation occur because catastrophic floods are generated in many cases by nonrandom occurrences of atmospheric phenomena (Ely et al., 1993, 1994, 1996). The scientifically interesting goal is to document such occurrences by developing long-term paleoflood datasets at multiple localities. Without such information it is impossible to evaluate the validity of any statistical flood frequency analysis, especially those based only on temporally restricted conventional datasets. Ultimately, the nonstationarity "problem" in flood frequency analysis will have to be viewed in the context of hydroclimatological models for long-period climatic variations. Such models hold the promise of meshing paleoflood studies and other aspects of flood geomorphology with the relatively short time scales of gaged flow records.

Opportunities

Independence, Stationarity, and Homogeneity

The USGS interpretive studies in regional flood frequency analysis present a timely opportunity for technology transfer in reevaluating traditional methods used in flood frequency estimation. Common practice estimates flood frequency by using annual or partial duration series of flood peaks. Various statistical estimators have been developed, all of which assume that historical flood peak series represent a sample of independent realizations drawn from a stationary homogeneous stochastic process. Emerging research provides growing evidence of deviations from these assumptions. A number of natural and anthropogenic sources of variability may result in nonstationarity in flood peak series. Secular trends in climatic variability can be demonstrated using traditional gaged records (Bradley, 1998) as well as through the use of dendrochronology, geomorphic interpretation of sediment deposits, and other paleohydrologic techniques (Jarrett, 1991; Enzel et al., 1993; Salas et al., 1994).

The gaged record of flood peaks at a site may frequently contain flood events caused by several very different mechanisms, such as rainfall and snowmelt, or tropical and nontropical storm systems (Hirschfield and Wilson, 1960) following varying antecedent watershed conditions. Similarly, systematic climate variability such as the El Niño/Southern Oscillation can result in predictable transitions between persistent climate states that exhibit significant and fundamentally different hydrologic characteristics (Dracup and Kahya, 1994; Canadian Electricity Association, 1994). The distribution of flood peaks under these circumstances may better be described as a mixture distribution of two (or more) distinct populations (Hirschfield and Wilson, 1960; Waylen and Woo, 1982) rather than a single homogeneous distribution. Systematic hydrologic variability (Landwehr and Slack, 1990; Cayan and Webb, 1992) can produce interannual persistence in hydrologic extremes that significantly alter flood frequencies and flood risks estimated from historical flood peak series (Webb and Betancourt, 1990).

The use of historical gage records and paleohydrologic techniques can similarly be used to identify persistence in hydrologic extremes affecting flood frequency estimation. Booy and Morgan (1985) have demonstrated interannual persistence (clustering) of flood peaks in Canadian streamflow records, using a Monte Carlo test for the Hurst phenomenon. Accounting for serial correlation in gaged flood peaks, the estimated recurrence interval of the design discharge for the city of Winnipeg's levee system decreased from 169 years to 70 years. The USGS's activities in regional flood frequency studies represent an opportunity to reevaluate traditional estimates of flood risk, accounting for improved understanding of the spatial and temporal characteristics of hydrologic extremes.

Alternate Methods for Estimation and Regionalization

The Survey's technical and institutional expertise may provide the opportunity to lead a systematic reevaluation of statistical methods for flood frequency analysis. The research advances that have been made since the adoption of the recommendations in *Bulletin 17-B* (USIACWD, 1982) have been significant. The Survey is strategically positioned to coordinate a systematic reevaluation of "standard" flood frequency techniques commonly used in the National Flood Insurance Program and the federal design and planning process by the Corps of Engineers and other agencies. Along with improved techniques for flood quantile estimation, the explicit computation of quantitative confidence limits represents essential information for planning, resource allocation, and risk-based decision making.

Extreme Events and Risk-Based Decision Making

Extreme event information commonly developed through interpretive studies (such as regional flood frequency analyses) also provides the information base to support improved risk-based decision making. For example, the observation of elevation effects in Rocky Mountain flood peaks allowed Jarrett (1993) to reevaluate flood frequencies, recognizing the significance of the linkages between the spatial patterns of orographic rainfall in regions of complex terrain and the associated flooding mechanisms (Barros and Lettenmaier, 1994; Barros and Kuligowski, 1998). This observation, supported by geomorphic paleoflood evidence (Grimm et al., 1995), dramatically lowered the estimate of extreme flood magnitudes in these high-elevation watersheds. While hydrologically interesting, this analysis clearly has direct implications in both the design and the operation of water resource systems in these watersheds (see Box 4.1). Similar integration

BOX 4.1
Elevation Limit to Rainfall Flooding

Prior to recent USGS studies it was widely believed that large rainfall floods could occur at any elevation in the Rocky Mountains. One such rainfall flood was the Big Thompson flood of 1976, which killed 140 people and caused over $35 million in damages. For more than a decade USGS hydrologists have documented the size of many contemporary and prehistoric floods on rivers throughout the Rocky Mountains. An analysis of this large detailed dataset by USGS hydrologists indicates that there is an elevation limit to rainfall flooding. Floods in river basins above about 5,500 feet in the Northern Rocky Mountains (or 7,500 feet in the Southern Rocky Mountains) are comparatively small and result from snowmelt rather than high-intensity rainfall. This research has led to substantially lower estimates of the 100-year flood (and the probable maximum flood) in high-altitude basins throughout the Rocky Mountains.

These results have important implication for floodplain management, implementation of flood warning systems, and the design of hydraulic structures in floodplains. Dam safety guidelines developed before this USGS research was done suggested that many dams in the Rocky Mountains were underdesigned. The cost of rebuilding spillways throughout Colorado to meet these guidelines was expected to be $184 million. One of the dams thought to have an underdesigned spillway is Olympus Dam in Estes Park, Colorado, located 7,500 feet above sea level. The spillway is designed for a flood of 22,000 cubic feet per second (cfs). The guidelines for spillways in this area would have required a redesign to accommodate a flood of 84,000 cfs. The USGS research showed that no floods in this basin had ever exceeded 5,000 cfs in the past 10,000 years. Thus, the USGS was able to demonstrate that the costly spillway reconstruction at this high altitude was not necessary, and spillway criteria for the Rocky Mountain region are being rewritten to reflect these findings. These downward revisions of flood risk mean that spillway modifications will not be necessary at some dams and that reservoir storage set aside for flood control can be used for water storage for municipal, industrial, irrigation, recreation, or habitat-related uses. Thus, the USGS findings not only result in a savings of redesign costs but also in more beneficial storage in the existing reservoirs.

of risk-based decision making and flood frequency analysis is observed in the ongoing discussion over the appropriate use of expected probability estimators and maximum likelihood estimators (NRC, 1995; Stedinger, 1996) to estimate flood exceedance probabilities and the likelihood of future flood damage.

Cumulative Impacts of Watershed Activities

Changes in catchment hydrologic response owing to land clearance (Bosch and Hewlett, 1982) and urbanization (Seaburn, 1969; Wallace, 1971; Ferguson and Suckling, 1991) have been well documented. The resulting nonstationarity in streamflow challenges the traditional assumptions and the use of historical streamflow records in flood frequency estimation. The regulation of large river systems (Collier et al., 1996) alters the flow duration characteristics of streams through both the storage and release of runoff and the resulting changes in channel form (Pizzuto, 1994). Nonstationarity from development activities represents a challenging opportunity to account for human-induced changes in estimating flood frequency.

The Hydroclimatic Data Network (Slack and Landwehr, 1992) represents a core set of stream gages selected to provide a representative database of unregulated streamflow throughout the nation. The ways in which land clearance and development affect existing streamflow records show the need for maintaining continuous records on relatively unregulated streams as well as refining techniques for flood frequency analysis that account for land-use changes and other sources of nonstationarity.

Paleohydrology

Recent developments in the field of paleoflood hydrology and statistical techniques for making use of such information provide new and often unappreciated opportunities for improving estimates of flood frequency relations at gaged and ungaged sites (O'Connor et al., 1994; Enzel et al., 1996; Ely, 1997; Baker, 1998). The USGS has over the last several decades collected streamflow records at thousands of sites in the United States. This information provides a vast resource for flood frequency and other investigations. Still, when frequency analyses are required at particular sites, often there is no streamflow gage located near that site requiring the use of regional estimates. In fact, even with a gaged record available, streamflow records are generally shorter than optimum to estimate the 50- or 100-year flood employed in floodplain planning, and the design of bridges, culverts, and flood control structures.

The availability of better statistical techniques has invigorated the development of techniques for obtaining botanical and physical paleoflood records, and suggested new and revised procedures and approaches for field investigations of historical, botanical and physical paleoflood information (Stedinger and Baker, 1987). The procedures for collecting this information in the field need to be better documented and made more widely known. Limitations on interpretation and accuracy of paleodata need to be carefully documented.

Role of the USGS

Interpretive studies provide consistent databases developed with uniform data collection techniques to support research issues. These efforts can, by design, include measurement of land-use and watershed characteristics to support current research and future interpretive analyses. The USGS also plays a crucial role in rapidly mobilizing resources to document and study extreme hydrologic events when they occur. This function is vital to the ongoing study of hazards from hydrologic extremes.

The USGS interpretive studies serve a national technology transfer function, providing cooperators with access to new techniques that support design and hydrologic risk management. The USGS is well qualified to provide national leadership in the development and application of emerging science in climate variability, improved statistical estimation, hydrologic regionalization, and the impacts of changing land uses on the evaluation of flood risks and risk-based decision making.

FLOOD/RIVER GAGING

State of the Practice

The USGS is the primary federal agency responsible for supplying hydrologic data to federal, state, and local agencies as well as private users. The USGS operates and maintains more than 85 percent of the nation's stream gaging stations, including 98 percent of those used for real-time river forecasting. The national stream gaging program is more fully described in Wahl et al. (1995).

One of the crucial roles of the USGS is to maintain current rating curves for stream gages used in forecasting and management of the nation's water resources. Changes in river cross-section geometry (Moody and Meade, 1990) can alter the stage-discharge relationship used to convert gaged stream heights into estimated discharges. The USGS routinely verifies and updates these rating curves and provides revised ratings to the user community. The agency also assures the quality and accuracy of archived historical streamflow data. During flood events, when scour, erosion, and out-of-bank flows can significantly change the cross-section of rated river channels, the USGS performs direct streamflow measurements to allow forecasters and managers to incorporate rapidly changing conditions in their management activities. For example, during the great flood of 1993 in the upper midwestern United States, over 2,000 field visits to collect supplemental discharge measurements or to check and repair gaging equipment were made by USGS hydrologic technicians (National Weather Service, 1994).

Faced with growing demands for hydrologic data in a time of limited resources, the USGS continues to develop and apply techniques intended to help optimize the national stream gage network through the use of new cost-effective measurement technologies and optimal network design (Medina and Tasker, 1985). Extensions of hydrologic regionalization can help quantify the value of adding or discontinuing gaging activities at a particular location. Gaging networks can therefore be designed, analyzed, and modified through the use of formal mathematical optimization techniques that balance the cost and information content of gaging activities (Moss and Tasker, 1991, 1995; Tasker, 1991b).

Opportunities

While the USGS maintains most of the stream gages in the nation, the National Weather Service, as the federal agency responsible for issuing flood forecasts, relies heavily on the accuracy and reliability of these gages for real-time river forecasting. Since most river gages record river stage rather than discharge, one of the significant sources of error in real-time flood forecasting can be the dynamic changes in mobile bed channel cross-sections during floods. Current practices address this problem through supplemental discharge measurements made during flooding events, when channel changes are believed to be significant. These supplemental measurements are not always possible during the particularly hazardous conditions associated with extreme floods (when they might be most valuable). The USGS expertise in scour and sediment transport, fluvial geomorphology, river hydraulics, and hydraulic measurement technologies may provide the basis for supplementing discharge measurements during floods with real-time process-based adjustments to gage rating curves.

Multiobjective Network Design

The USGS's expertise and application of optimization techniques to the cost-effective design of stream gaging networks provides a methodological foundation to help balance the competing demands for hydrologic data that are increasingly placed on the national stream gage network. For example, long records of unregulated discharge, such as the Hydroclimatic Data Network, provide an invaluable database to support research and risk-based decision making related to climate variability and change. In contrast, flood forecasting to protect lives and property may demand gaging for critical catchments associated with widespread land clearance and development. Growing demands placed on the national gaging program suggest opportunities to employ multiple-objective optimization in the design and modification of gaging networks. These techniques may be usefully applied to consider network design with mixed sensors (such as continuous and peak stage recorders) as well as tradeoffs between potentially competing demands for limited gaging resources (e.g., flood vs. drought needs, or flood warnings vs. climatic monitoring).

New Measurement Technologies

The National Research Council (1992) has noted the need and opportunity to take advantage of new and emerging technologies to support the national stream gaging program. The National Weather Service's deployment of the WSR-88D

Doppler weather radar was identified as one such timely opportunity. The evolving technologies in dual polarized radar (see Box 4.2) represent an extension that may ultimately be added to the WSR-88D weather radars.

Obtaining data, particularly in an era of funding cutbacks, may require new and innovative instrumentation and monitoring techniques. Recent advances in microprocessors, electronics, and satellite communications are leading to a new generation of reliable, precise, and relatively inexpensive field instrumentation equipment. Improved instrumentation for stream gaging may improve the economy and reliability of the nation's gaging network for flood hazard evaluation.

Large-scale remote sensing capabilities, such as NASA's Mission to Planet Earth and defense technology conversions, afford many opportunities to enhance the study of extreme events through cooperative research initiatives and interagency coordination. Existing and anticipated multisensor earth observations from space can document inundated land areas, suspended sediment concentrations, basin characteristics, and even approximate peak discharges for individual extreme flood events. When combined with local gage information and postflood field surveys, such monitoring from space can supplement conventional databases on extreme floods, presumably at greatly reduced cost.

In addition to more accurate and cost-effective technology, the diverse needs of cooperators suggest the value of multiple levels of reliability in stream gaging. As resources for data collection, processing, publication, and archiving have

---

BOX 4.2
Improved Rainfall Estimates with Modern Instruments

Modernization and restructuring of the National Weather Service included the deployment of a national network of WSR-88D Doppler radars, which estimate precipitation rates from radar reflectivity. These estimates of precipitation rates can be integrated over time to yield estimates of precipitation amount. While the spatial and temporal coverage is excellent, approximately 2 kilometers and 6 minutes, respectively, there are accuracy problems with the precipitation amounts, especially in snow and convective situations.

Field experiments with alternating horizontally and vertically polarized radar signals have been successful in assessing the size and shape of falling precipitation particles. This "dual polarized" radar yields vastly improved estimates of rainfall rates. There is a distinct possibility that further testing will lead to an upgrade of the existing WSR-88D Doppler network to include dual polarization capability. The resultant rainfall data have the potential to revolutionize rainfall-based hydrology as we know it. Accurate rainfall estimates would be available over most of the continental United States every 15 to 30 minutes with an average horizontal spacing of 2 kilometers. Runoff and other hydrologic models will require extensive rethinking to take full advantage of this improved dataset.

declined, some users have elected to support only the maintenance of real-time measurement and telemetry capability at existing gage sites, without incurring the additional expense to verify and publish provisional observations. For water managers primarily interested in real-time river stages (e.g., for real-time reservoir operation), this may represent a cost-effective tradeoff. Alternatively, the USGS may want to consider offering cooperators data with lower levels of reliability, at lower cost. For example, for water managers primarily interested in accurate measurement and monitoring of low flows, cost-effective telemetered sensors are available that can be inexpensively installed on existing structures such as bridge abutments and river intake structures. While these installations would not be expected to survive extreme flooding, the sensors could be removed seasonally, or replaced if lost due to flooding. Compared to the expense of constructing and maintaining a powered gage house providing standard levels of reliability, this may offer a cost-effective data source for many users. The integration of real-time data collected by state and municipal data collection networks may similarly represent a useful supplement to the Survey's core gaging network. Synthesis of data from a variety of sources may require little more than the cost-effective agreement on common data standards between agencies.

Role of the USGS

The provision of reliable high-quality streamflow data remains a fundamental and invaluable role of the USGS. Interpretive studies directed at improving the effectiveness of the nation's stream gage network will continue to support this vital mission as the demand for streamflow data continues to experience resource limitations. The USGS's role in continuously improving the design and operation of the national gaging program will remain essential, even as new technologies and interdisciplinary hydrologic methods are integrated into USGS activities.

RIVER SCOUR

State of the Practice

Investigative studies of bridge scour conducted with local and regional cooperators provide a consistent information base to support research on scour and sediment transport. Interpretive studies also provide the mechanism to transfer products of the USGS research effort to other federal agencies, cooperators, and resource managers. Scour studies addressing the risk of failure for bridges and highway structures have a more direct risk management focus than flood frequency estimation. Beyond developing design criteria, the application of rapid screening techniques (Holnbeck and Parrett, 1997) provides a cost-effective means for resource managers to target detailed scour investigations for structures posing the greatest risks. Investigative studies also provide case studies of failure modes for forensic analysis, such as the 1995 failure of the I-5 bridges across Los Gatos Creek in California. These investigations support verification of analytical techniques and suggest additional research priorities.

Opportunities

Products of the National Research Program have suggested criteria for risk-based design of bridge and highway structures that can be more cost effective than traditional design criteria. The USGS's interpretive studies on bridge scour provide the opportunity to integrate risk-based criteria in the design of vulnerable structures. Scour-related failures of bridges and highway structures represent opportunities to better understand the critical factors and potential deficiencies in design criteria and estimated scour risks. The evaluation of failures may be enhanced by distinguishing failures in which the original science or design data may have been inadequate or failures due to hydrologic changes or errors in estimating design conditions. Adjustments of flood frequency estimates resulting from nonstationarity and persistence can result in adjustments to scour-based designs as well.

Scour on Floodplains

Severe flooding on the San Jacinto River in Houston resulted in scour damage to several major interstate pipelines transporting refined petroleum products along the eastern seaboard. The scour-related failure included scour in the broad coastal plain floodplain, resulting from major out-of-bank flows. Estimates of scour depth are used in traditional design practices for pipeline river crossings. Deep scour in the lithified materials of the coastal floodplain represents a low-probability/high-consequence event, for which protection in the form of deep excavation is not generally considered cost effective. This analysis is constrained by the uncertainty associated with estimating scour risks in floodplains. The need for risk assessment and cost-effective mitigation from this hazard represents an opportunity to extend the USGS's expertise in flood hydrology, geomorphology, and scour mechanics to develop management criteria for this low-probability/high-consequence hazard.

Nonstationarity

The cumulative impacts of watershed activities can significantly alter the sediment budget and flow frequency characteristics of rivers (Graf et al., 1991), resulting in significant changes in scour risks. Beyond the use of new scour equations, the appropriate design flow selected for rivers experiencing, or expected to experience, significant land-use changes, represents a risk-based allocation problem balancing higher design and construction costs against the risk that design discharge quantiles will change in the future. Interpretive studies of scour and failure in watersheds that have experienced land-use changes provide the opportunity to integrate research in watershed processes and the impact of land-use changes to risks and costs associated with the failure of structures located in the river channel.

The USGS's interdisciplinary expertise in flood frequency analysis, fluvial geomorphology, and sediment transport provides the foundation to identify the limits of current practice and identify new challenges and research needs in design criteria and risk evaluation. The expertise and measurement technology developed to support the scour program may be directly transferable to support the dynamic evaluation of discharge and channel cross-section changes during flooding, supporting the national gaging program as well.

Role of the USGS

Interpretive scour studies support improved scour modeling, prediction, and design criteria for bridges and highway structures. The innovative use of acoustic Doppler profiling dramatically expands the information base on channel dynamics under high-flow conditions, supporting refinement of methods for scour prediction and stable design. Regional interpretive studies conducted in partnership with the states represent both technology transfer mechanisms from the USGS research program to local cooperators, and opportunities to verify and evaluate improved management and prediction capability.

DROUGHT FREQUENCY AND HAZARD

State of the Practice

Frequency estimation techniques developed for floods are also applied to the estimation of low-flow frequencies (Vogel and Kroll, 1990). Regionalization of low-flow characteristics supports low-flow frequency analysis at ungaged sites. The USGS also conducts studies directly related to the management of drought risks (Hirsch, 1978), as well as the potential drought impacts of climate change (Tasker, 1991a, 1993; Tasker et al., 1991; Wolock et al., 1993).

Opportunities

Frequency estimation for low flows raises many of the same issues related to assumptions of stationarity, homogeneity, and independence encountered in flood frequency estimation (Booy and Morgan, 1985; Cayan and Webb, 1992; CEA, 1994). The estimation of extreme drought risks represents a significant opportunity for both technology transfer from the National Research Program and integration of hydrologic science with planning and resource management. While existing gage records provide data for the application of traditional frequency estimation techniques to drought risk, longer secular trends in streamflow may alter the drought risk estimated from the relatively short database of gaged streamflow. For example, tree ring evidence has suggested the unusually wet nature of the available hydrologic record that served as the basis for the "over-allocation" of the Colorado River in 1922 (Stockton et al., 1985). Dendro-climatogy used by Cleave and Stahle (1989) similarly indicated significantly greater interannual persistence in hydrologic extremes than that recorded in twentieth-century stream gage records on the White River of Arkansas. A substantial opportunity exists to improve the accuracy and to quantify the uncertainty in estimates of hydrologic extremes under nonstationary conditions.

While the western United States experienced an extreme multiyear drought from 1985 to 1991, no comparable drought has been recorded in historical streamflow records in the humid east. A number of Atlantic slope basins nevertheless experienced persistent droughts in the 1930s and 1960s. Given this experience, prudent water management would attempt to determine the likelihood of a "California" drought in the east. Traditional frequency analysis based on the assumed independence of annual low-flow series cannot provide an adequate response. The USGS's expertise in hydroclimatic linkages and flow frequency analysis represents an opportunity to improve the accuracy and quantify the uncertainty in estimating extreme drought risks.

The spatial extent of extreme droughts may substantially exceed boundaries of hydrologically "homogeneous" regions identified with traditional regionalization techniques. For example, the drought of 1988 was rare not only in the magnitude of the precipitation deficit and low streamflows recorded at sites throughout the nation but also in the spatial extent of drought conditions across the midwest (Kunkel and Angel, 1989; Kunkel et al., 1989). (The Great Flood of 1993 manifested similarly extreme in both the magnitude of precipitation—about 30 inches in 4 months—and the spatial extent of summer flooding (Lott, 1993).) These events suggest a varying spatial scale associated with the recurrence interval of extreme hydrologic events that may not be well represented using standard methods of hydrologic regionalization and frequency analysis. Moreover, these extreme events are least likely to be well represented in our relatively short historical streamflow records. The use of paleohydrology to evaluate the magnitude and frequency of hydrologic extremes, as well as the spatial extent (Jarrett, 1990) of extreme hydrologic events, represents a significant opportunity to integrate traditional regionalization approaches and the emerging understanding of hydroclimatic linkages to improve the estimation of extreme hydrologic risks.

Biological Versus Hydrologic Flow Duration

Frequency-based estimates of low-flow quantiles (such as the 7-day, 10-year low flow or 7Q10) are regularly used in water quality regulation to derive wasteload allocations and total maximum daily loads. The U.S. Environmental Protection Agency (EPA) also supports the use of "biological" low-flow quantiles, complementing the familiar statistical low-flow estimates (USEPA, 1990, 1991). Estimation of biological low-flow quantiles has been developed to reflect the continuous recovery periods thought to be required by biological populations recovering from low-flow stress. Heuristic techniques have been developed that utilize the historical hydrologic record to identify flow levels that would empirically achieve nominal frequency duration criteria, based on historical streamflow records. Biological low flows for acute, 1-hour/3-year (1B3) and chronic 4-day/3-year (4B3) low-flow quantiles are recommended and have been utilized by EPA (CFR, 1992) in setting toxic discharge standards for aquatic life protection. The uncertain relationship between biological and hydrologic flows (e.g., 7Q10) represents an opportunity to clarify management criteria and permit decisions through the integration of Water Resources Division's expertise in hydrologic frequency analysis and Biological Resources Division's expertise in habitat requirements for aquatic species. The Survey's strengths in these areas suggest a natural opportunity to integrate research on low flow frequencies and habitat requirements with the science-based regulatory initiatives in instream flow maintenance and wetland restoration within the USEPA and USDA.

Role of the USGS

Drought-related interpretive studies expand the information base supporting research (Liu and Stedinger, 1991) and represent a valued mechanism to transfer research products to resource managers and cooperators (Ludwig and Tasker, 1993). The USGS continues to play an integral role in support of drought management, serving as a source for interdisciplinary hydrologic expertise.

5

Communicating Information on Hydrologic Hazards

The U.S. Geological Survey's (USGS) mission includes providing reliable, impartial, and timely information needed to understand the nation's water resources and related hydrologic hazards. The devastating consequences of extreme hydrologic events have caused some to call for the agency to expand its role from providing data for decision making to improving the understanding of hydrologic risk. Former Director Gordon Eaton observed that "USGS employees must make connections necessary to establish a new compact between the earth sciences in the society we serve. It is up to us, then, to ensure that our work is understood and applied" (Gordon Eaton, personal communication, benchmark letter of March 27, 1995, to all USGS employees). In response, the USGS mission statement encourages efforts that "actively promote the use" (Robert Hirsch, personal communication in briefing to committee 1997) of the information the agency generates.

For hydrologic hazards, "outreach" refers to activities through which USGS communicates information about the probability and consequences of natural hazards. Outreach efforts improve "customer service" and raise the general public's awareness of USGS expertise and activities having a critical and significant connection to their own lives. In operational terms the purpose of outreach is to effectively communicate to decision makers the nature, probability, and consequences of hydrologic hazards. With this understanding individual and institutional decision makers in the public and private sectors can make better-informed investment and regulatory decisions. Enhanced understanding of hydrologic hazards in the service of better decision making is one standard by which the success of the USGS's outreach efforts should be measured.

OUTREACH GOALS

The USGS can focus its outreach program by continually measuring outreach requests against a carefully defined outreach mission: *improving understanding of hydrologic hazards so that better investment and regulatory decisions can be made by private and public interests.* This mission focus may direct the USGS away from general educational products toward ones that support decision making. Nonetheless, the USGS mission in outreach should not be to direct particular decisions about hydrologic hazard management (e.g., particular floodplain development patterns). Instead the agency should assure that decisions are made with the best possible understanding of the probabilities and consequences of hydrologic hazards.

This focused outreach mission might better be described as helping decision makers avoid being "surprised." An informed decision maker who understands the probabilities and consequences of drought or flood hazards can make educated choices and will not be confronted with unanticipated outcomes when a drought or flood occurs. Similarly, relevant understanding of the physical processes and real-time risks associated with hydrologic hazards enables emergency managers to make informed decisions that minimize the threat to lives and property during hazardous events.

HISTORICAL DEVELOPMENT OF USGS OUTREACH

The role of the USGS in providing reliable and impartial information on hydrologic hazards has changed dramatically in the past 30 years, driven by a changing customer base and rapidly evolving information technologies.

The primary customers for hydrologic data and related information in the first two-thirds of the twentieth century were the agencies and organizations responsible for developing the nation's water resources infrastructure. The U.S. Army Corps of Engineers (USACE) and Bureau of Reclamation provided funding for the USGS to collect hydrologic data needed for the planning and design of all the major reservoirs and related flood control, navigation, water supply, and power generation facilities. State highway departments also provided funding through the Federal-State Cooperative Investigations Program to collect data and provide information on flood characteristics of small streams to support safe and economical design of bridges and culverts. The products that USGS provided during this period were the historical quality-controlled streamflow data and some analyses of data related to water supply risks, reservoir safe yields, and flood frequency determinations. The National Weather Service (NWS) installed instruments in many USGS stream gaging stations for transmitting river stage via landline to their forecast offices. USGS did not transmit real time data directly to any organization. During this period USGS's primary outreach activities were

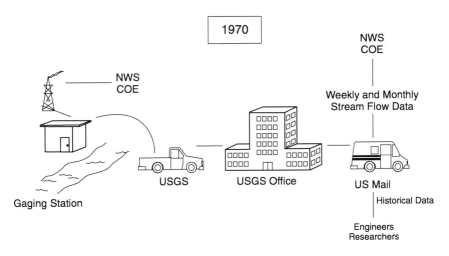

FIGURE 5.1 USGS's mode of data collection and dissemination in 1970.

the provision of reliable historical streamflow data and local interpretive information on historical hydrologic risks.

The USGS role began to change in the 1950s and 1960s as streamflow monitoring expanded to support reservoir operations, water quality monitoring, and detailed flow monitoring associated with the allocation of limited resources, particularly in the western states. Much of this information was needed in real time to support decisions on water releases from reservoirs and withdrawals for municipal supplies, and to maintain ambient water quality within permit requirements. The USGS role began to expand during this period to include the provision of limited real-time data to agencies with specific real-time management needs. A schematic of the typical mode of data transfer to USGS's customers in 1970 is shown in Figure 5.1. Reliable historical streamflow data during this period were primarily provided in printed form, with provisional real-time data available to specialized users. Interpretive information on historical hydrologic risks became more widespread and standardized through USGS's flood studies and flood frequency analyses.

In the 1980s and 1990s, the traditional customer base for USGS expanded significantly. Organizations providing new funding to maintain and operate the stream gaging program included city and county agencies that needed information for many different purposes, including flood hazard warning. Today, more than 800 agencies support the stream gaging program and at least two-thirds of them are local government agencies. This change in USGS customer base coincided with dramatic improvements in the quality and cost effectiveness of satellite and microprocessor technologies available for data management and dissemination. Instead of recording data on paper charts or punched paper tapes, USGS

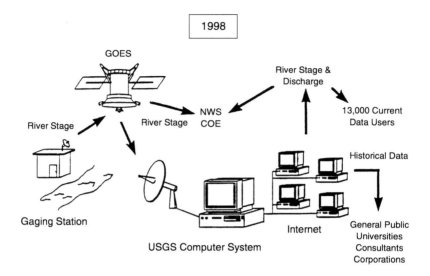

FIGURE 5.2 State-of-the art mode of data collection and dissemination for USGS in 1998.

began storing data on small storage modules, downloading data to field computers for automatic processing, and transmitting data from field sites via the GOES satellite network, as shown in Figure 5.2. The satellite capability enabled USGS to transmit, process, and disseminate both river stage and discharge data every few hours to its customers who needed the information for emergency warning and response functions. Examples of flood data being transmitted directly to customers include King County Emergency Services in Washington, the New Jersey State Police, and the Louisiana Department of Emergency Services. River stage data also became available to other customers, such as the NWS and USACE, with access to the GOES satellite information.

FUTURE USGS OUTREACH

Continuing expansion of the USGS customer base, along with accelerating improvements in information technologies, can be expected to propel USGS's outreach efforts in hydrologic hazards in new directions. Growing demands for expanded hazard information can be expected to be driven by increasing activities to actively manage local risks associated with hydrologic hazards, expanding awareness of hydrologic hazards and their potential for catastrophic losses among the general public, and continued growth in properties and populations at risk from hydrologic hazards. In planning to support continuous improvement in hydrologic hazard outreach, the USGS must be able to match its unique scientific expertise and data management capabilities to the evolving needs of an expanding user community. This challenge will require thoughtful coordination to balance competing demands for limited resources with the needs to demonstrate the value and cost effectiveness of outreach efforts. The evolution of traditional USGS outreach activities in response to new technologies and information needs can be expected to grow in three primary areas: (1) synthesis of USGS expertise in science and fundamental processes to characterize historical (or unconditional) hydrologic risks, (2) dissemination of provisional hydrologic data in real time, and (3) data and information support of emergency management and disaster response agencies.

HYDROLOGIC HAZARDS OUTREACH

Historical Risk

The USGS maintains unique capability to synthesize historical data with understanding of fundamental physical processes and geoscience, in order to characterize the historical risks of hydrologic hazards. In this context historical risk refers to unconditional risk information that is not dependent on current hydrologic conditions, such as observed river stages or reservoir levels. Examples of traditional USGS activities in characterizing historical risks include flood frequency analyses and floodplain mapping. The Survey also provides essential data collection and data dissemination functions to support the management of real-time or conditional risks. The analysis, communication, and management of real-time risks, such as real-time flood forecasting or flash flood warnings and evacuations, are primarily the responsibility of operational emergency management agencies, indispensably supported by real-time data provided by the USGS.

The unique strength of the USGS hazards activities in geoscience process studies and modeling provides improved understanding of the likelihood and consequences of hydrologic phenomena. Process studies and models have been developed as contributions to basic science. The USGS publishes its research results in the professional literature and provides technical report to decision makers. This basic science can provide the foundation for additional creative outreach efforts, but these publications alone may not satisfy the broader emerging demands for decision-relevant risk information on hydrologic hazards.

The evolving goals for successful hazards outreach include improved understanding by decision makers in addition to fundamental contributions to scientific understanding. Such understanding begins with basic science but also requires new approaches to effectively communicate that science. Scientific and technical papers alone may not be sufficient to adequately convey the emerging understanding of flood and drought phenomena to the USGS's broad audience of decision makers. The explosive growth in graphical and computational capability provides powerful tools to help decision makers better visualize the probabilities and consequences of hydrologic hazards, in a context relevant to their specific decision problems. The committee sees rich, timely, and cost-effective interdivisional opportunities for the USGS to integrate resources in its mapping program with its interdisciplinary scientific expertise in fundamental process studies, modeling, and data collection, in order to substantially improve outreach activities that communicate the historical risks associated with hydrologic hazards.

Real-Time Data Management

Rapid growth in the use of the Internet has greatly expanded the USGS's potential audience and approach to disseminating hydrologic data. Potential USGS customers are no longer restricted to government agencies with operational or emergency management responsibilities; customers now may include individual property owners, recreational boaters and fishermen, and virtually any organization or individual with interests in current streamflow data. Use of the Internet for dissemination of both real-time and historical data has expanded rapidly. The USGS is currently providing real-time data on the Internet for more than 3,900 stream gaging stations, and this number will continue to grow. The agency is a national leader in the use of the Internet for these purposes. The USGS programs for data are impressive, and the agency should continue to expand these efforts.

For the USGS, widespread dissemination of provisional data in real time is a relatively new and rapidly developing outreach product. Real-time data, in conjunction with historical risk information, create new opportunities for real-time hazard management.

Real-Time Hazard Management

Emergency response refers to the ability of individuals and local emergency management agencies to take actions that will minimize the adverse consequences of hydrologic hazards. One of the most common and widespread benefits of such actions is reduced loss of life and property damage through enhanced warning and coordination of evacuations. In more extreme circumstances decisions might be made to intentionally breach a levee or modify reservoir operations. The USGS should continually seek out opportunities for improved transfer of data and use of these data for emergency response by both individuals and governments.

The USGS is the primary reliable source of real-time streamflow data to support flood forecasting by the National Weather Service and state and local emergency management agencies. Real-time access to provisional streamflow data is also available from a growing number of stream gages through the Internet. One of the challenges in exploiting the growth in data availability afforded by these new information technologies is to foster the simultaneous growth in decision makers' *understanding* of the information derived from these data. Despite exponential growth in the use and availability of the Internet, it is essential to recognize that real-time data dissemination to support emergency response under life-threatening conditions requires a level of reliability and redundancy unavailable through the Internet. A second challenge in supporting real-time hazard management is to expand the real-time transmission of data via GOES satellite, VHF radio, telephone, and the Internet to establish reliable and redundant data dissemination avenues, including commercial radio and television.

Reliable dissemination of hydrologic data to the broader public is a critical mission application that cannot be undertaken by USGS alone. In close collaboration with local emergency managers, the USGS experience in supporting the real-time gage network provides the foundation for broad-based cooperative hazard management.

Improved public access to real-time data will be enhanced by closer relationships with radio and television media. Outreach for real-time hazard management will require close coordination by the USGS (as the provider of data), the local and regional emergency management agencies, and the news media. This evolving mode of real-time data distribution may require USGS's outreach efforts to include participation and leadership in training and preparedness exercises, in order to support and coordinate timely and accurate dissemination of critical real-time data during infrequent flood events. Simply making the data available in real time will not result in their timely and effective dissemination. The expanding USGS role in these activities should be accompanied by programmatic monitoring and continuous evaluation in order to assess the value, effectiveness, and long-term success of outreach efforts supporting real-time hazard management.

SUPPORT FOR RISK COMMUNICATION

Risk Perception

Federal agencies involved in project construction, in establishing rules for selling insurance, and in the distribution of disaster aid for postflood reconstruction have difficulty communicating the nature of flood risks. These agencies, along with the USGS, have a sophisticated understanding of flood risk that may not be easily communicated to local decision makers.

Conveying an understanding of the concept of a flood recurrence interval remains an elusive goal. The expression "100-year flood" was developed as a useful shorthand for professional communication. Yet as the term moved into popular use it became misunderstood and is commonly misinterpreted as a flood that will only occur once every 100 years. Suggestions have been made to replace the term with a percentage chance of occurrence (e.g., the 1 percent chance flood). Yet we know from the behavioral sciences that, as the likelihood of hazardous events decreases, individuals tend to truncate the perceived probability distribution, effectively acting as though a low-probability event has a zero likelihood of occurrence (Slovic, 1977; Schoemaker and Kunruther, 1979; O'Grady and Shabman, 1994). One suggestion has been to communicate low probabilities in different terms; for example, an annual probability of a home being flooded instead might be presented as the chance of the home being flooded at least once during the period of a 20-year mortgage. Effective risk communication is a continuing challenge in hazards outreach. New nontraditional approaches to communicate probability concepts represent timely opportunities to improve risk communication in USGS outreach efforts for hydrologic hazards.

Residual Risk

In choosing the structural reliability of projects, as well as the degree of hazard reduction that is affordable, local interests and federal agencies need to fully appreciate the hydrologic hazard remaining after project construction. Nonstructural hazard reduction, such as the sale of flood insurance linked to delineation of the 100-year floodplain, similarly creates the impression of regions that are hazardous and complementary regions that are safe. Since these areas remain subject to flooding, there is a residual risk to any new development located in these areas. (This problem was highlighted in a recent report of the National Research Council on the American River Basin in California (NRC, 1995). Risk communication to improve decision making must therefore include the representation of residual risk.

Alternate Means of Communicating Risks

There are a number of opportunities for the USGS to pursue in the interest of communicating information about flood risk and consequences. USGS could tie together computer graphics and visualization capabilities (drawing upon geographic information system and mapping technology) with the agency's understanding of fundamental hydrology and its extensive databases. The objective would be to use the new technologies to visually simulate the sequence of flood events at particular locations as a way to communicate hazards at those locations. Such a tool would be of immediate use to local zoning officials, insurers, homeowners, and potential home purchasers as well as those responsible for making public investments in flood control works. USGS activities integrating digital mapping and hydraulic modeling for floodplain analysis are readily presented in such a visual form.

The various consequences of levee building can provide a good illustration of how such a tool might be used. In rivers not bounded by levees the effects of overbank flow diminish with distance. In locations where levees protect the floodplain, if flows overtop the levee there is a discontinuity in the relationship between flood flow and stage and its impact on floodplain structures and inhabitants. If a levee fails, the sudden rush of water may inundate structures by many feet of water and cause more damage than would have occurred if the levee had not been constructed; a levee failure is akin to a flash flood.

Efforts to convey the consequences of these different situations are typically made in tables of numerical values. For example, lives at risk might be estimated or the monetary value of the potentially inundated property might be listed. However, such tables of data may not adequately convey flood consequences or the character of the area that would be inundated. Floodplains surrounded by rivers and having a bowl shape present a different hazard than land that slopes up as one moves away from a river. How deep would be the water be? How much warning would there be? Could people escape? It is important to tailor assessments of risk to the specific features of a watershed or floodplain.

One way to add diversity to descriptions of flood risk is to have decision makers create realistic scenarios. The cooperative building of scenarios can be an excellent way to communicate flood risk. For example, the following scenario was suggested by the NRC as a possible description of the vulnerability of the Sacramento and Natomas, California, areas for storm events that overtop the levee system in the Flood Risk in the American River Basin (NRC, 1995):

> Should levees protecting Sacramento south of the American River be threatened, residents could attempt to move to higher ground to the south and west farther away from the river, and the depth of flooding would generally not exceed that at the river's edge; few areas would experience flooding of more than 10 feet. Natomas, on the other hand, is ringed by levees so that residents trying to leave the area would have to find their way across the main highway system to areas with higher ground that are primarily to the west. Moreover, because Natomas is in a depression, a third of the area would flood to over 10 feet and some to as much as 35 feet in depth. If the Natomas area is subject to a 1 in 100 chance of being flooded in any year, then the probability of at least one flood in 50 years is 40 percent. Therefore, the probability of a relatively catastrophic event within the lifetime of most residents is roughly equal to the probability of flipping a fair coin and getting heads.

Modern computer graphics could provide dramatic simulations depicting in time-lapse sequences a visual image of the area and the inundation flows. The results of different levee scenarios could then be developed and represented.

KEY OUTREACH ELEMENTS

Successful outreach for hydrologic hazards must begin by identifying the audiences for outreach. The growing diversity of audiences for hazards outreach will require the USGS to carefully match the demands for hazards information to both the specific nature of the hydrologic risks and the expertise the agency has to offer. In crafting this thoughtful structure, outreach efforts will also need to balance competing demands for the limited available personnel, graphics, publications, and other interpretive resources. To meet these challenges, outreach efforts need to be well conceived, identifying clear goals as well as unambiguous performance measures that will support the evaluation of outreach effectiveness. Within this delicate balance a broad spectrum of creative and innovative outreach programs can be cultivated.

Outreach at Cascade Volcanoes Observatory

USGS research at the Cascades Volcanoes Observatory (CVO) is intimately linked to the hazards associated with volcanic eruptions and the resulting risks to populations living nearby. The principal risk to the inhabitants of these regions is posed by lahars or volcanic mud flows that travel tens of kilometers from their sources and inundate everything within their paths. Yet, as the Pacific Northwest grows, development and population expansion persist in areas previously inundated by mud and debris flows. For instance, the area buried by past lahars from Mount Rainier now supports approximately 100,000 people.

The outreach activity at CVO was promoted, in part, by a growing awareness at the state level that natural hazards assessments were needed for high-growth-rate counties. Moreover, from 1986 to 1993, there were 23 glacial outburst floods in Mount Rainier National Park, and during one summer 20 cars were trapped by such an event. The high public awareness of hydrologic hazards defined a broad audience for outreach. This audience included educators, emergency managers, and coordinators with electronic and print media. Most of the products produced for this outreach effort are targeted to educators, students, and the general public and include educational materials, fact sheets, and exhibits. The project is conducted in cooperation with scientific staff from USGS and other regional agencies and research institutions, as well as illustrators and reports staff members at the USGS.

While the demands for hazard information at CVO were substantial and broadly based, the nature of the hazard required an unusual combination of expertise that was uniquely found within the USGS. Moreover, the specialized nature of the hazard provided few familiar analogs, and therefore required significant new tools and widespread efforts to effectively communicate hazard information. These characteristics led to an unusually vigorous and sustained outreach effort. This effort matched USGS expertise to broad demands for hazard information about historical risks, from an unusually diverse outreach audience.

Outreach in Louisiana Flood Warning

With 20 percent of all flood insurance claims and over $200 million spent annually in flood damages, Louisiana leads the nation in property damage caused by floods. The principal cause of flooding along the Amite and Comite rivers and their tributaries is backwater flooding. To alleviate and help reduce future losses, the USGS, in cooperation with the Louisiana Department of Transportation and Development, Louisiana Office of Emergency Preparedness, East Baton Rouge Parish, Amite River Basin Drainage and Water Conservation District, National Weather Service, and U.S. Army Corps of Engineers, operates a real-time flood monitoring and flood warning system in the Amite River Basin. The network uses satellite, VHF radio, and modem telemetry to relay data to the USGS and on to cooperators, the media, and the general public via facsimiles, the Internet, and recorded voice messages.

In contrast to outreach at the Cascade Volcanoes Observatory, the scientific expertise in the hydrology and hydraulics of Mississippi River flooding is not confined to the USGS. Many agencies with technical and operational expertise in disaster management, flood forecasting and operation, and emergency response participate in this collaborative endeavor in real-time hazard management. The common need among these participants is reliable real-time data. This need is well served by USGS expertise in river discharge measurement and the collection, management, and dissemination of hydrologic data, particularly under severe flooding conditions. The primary audience for this real-time data is the diverse community of public-sector management agencies with shared responsibility for risk interpretation and communication to the general public.

The USGS has also served as the focus for a complementary initiative to provide interpretive information to the general public in the form of a regional flood tracking chart. Developed on a map of the basin, the chart graphically depicts the five highest historical flood stages at each of the NWS forecast points in the basin. Used in conjunction with knowledge of historical flood limits and river forecasts disseminated by the NWS, this tool provides a simple method to help individuals interpret the consequences of forecasts and estimate the risks of flooding in the vicinity of gaged locations.

This effort matched USGS expertise in river gaging and hydrologic data management to the needs of regional emergency management agencies. Regional cooperative efforts allowed this expertise to be leveraged with the institutional forecasting and response infrastructure already in place. The critical nature of real-time data for decision making was served through the creative development of redundant channels for data dissemination, including the range of data transmission technologies and the popular media. This targeted well-defined data dissemination role was complemented by the joint development of an informational tool, targeted to a much wider nonspecialized audience, in the form of a regional flood tracking chart.

BALANCING RESEARCH AND OUTREACH

In both of these examples the information demands from the outreach audiences were matched to the unique expertise found within the USGS. Outreach efforts were effectively allocated between scientific, interpretive, and technical staff to address clear, well-defined, project-specific goals. These examples illustrate the value of targeted well-conceived outreach efforts to support the provision of reliable, impartial, and timely information needed to understand the nation's water resources and related hydrologic hazards.

The relevance of USGS hazards research will continue to intersect a growing audience for hazard information. Upon identifying audiences for its outreach program, the USGS may find itself receiving more requests for outreach products than it can readily provide. This is an expected outcome of a program that has high quality but does not charge for its services. The potential demand for data, analysis, interpretation, and direct educational support could easily exceed agency resources. Thus, the USGS mission in outreach should not be to direct particular decisions about hydrologic hazard management (e.g., particular floodplain development patterns or water conservation plans to address drought). Rather, USGS must focus on scientific principles in support of sound policies—not particular project specifications or management decisions. The agency's growing challenge will be to continually refine the balance between the expanding demands from outreach clients and the essential need to maintain its uniquely credible role as the nation's principal source of impartial expertise in geoscience information.

The resource demands to support hazards outreach may also require a re-evaluation of the allocation of effort between research, technical, and interpretive personnel within USGS. Coordination between the National Research Program and the district staff's direct customer contact will benefit from consistent incorporation of outreach efforts in project formulation, planning, and budgeting. The highly interdisciplinary nature of hydrologic hazards research and risk communication will require close integration of the divisional expertise that has evolved within the USGS, approaching outreach efforts as a single thread woven through the fabric of research and interpretive studies in hydrologic hazards.

This creative balance will require USGS to continually evaluate the most appropriate means to match its expertise to the needs of the expanding audiences for hydrologic hazards information. Although the agency has unique expertise in the earth sciences, expertise in outreach and risk communication has evolved within diverse organizations across many disciplines. USGS should consider a benchmarking process to cull the best of lessons learned from organizations performing outreach and communications activities. The encouragement of innovative approaches for outreach should be accompanied by thoughtful identification and evaluation of measures of success reflecting the balance between agency expertise, resources and priorities, and the diverse and specialized needs of the target audiences.

CONCLUSIONS

Previous reports by this committee have recommended that the USGS expand its educational efforts, but those calls have primarily been to encourage greater involvement in the education of earth and environmental scientists and increased research collaboration with universities and other institutes. These recommendations stemmed from the concern that, in order for the USGS to continue to pursue its data analysis and research mission, the agency needs to make certain that there is a cadre of professionals educated appropriately from which to recruit. In general, this report urges the USGS to take a broader view of education. Within the context of hydrologic hazards, the committee encourages the agency to establish outreach connections beyond its intellectual community to the general public in an effort to actively engage the support of informed decision making in the protection of life and property.

The role of the USGS has changed dramatically over the past 30 years. A number of factors, such as modified funding allocations within the federal government and the agency itself and the changing nature of environmental problems, have contributed to this transition. One outcome is that the USGS now has a very different set of partners and customers than in previous times. The other major factor that has stimulated many changes in the agency is the revolution in information technologies and the wide use of the Internet. Where a primary occupation of USGS had been to collect, analyze, and disseminate hydrologic data to other government agencies, researchers, and design engineers, now USGS has the capability to transmit real-time data to emergency managers and local citizens.

In recent years, then, the emphasis of the agency's mission to provide the nation with reliable and impartial information about the earth, to minimize the loss of lives and property from natural disasters, to manage resources, to protect ecological and human life, and to contribute to wise economic and physical development has shifted from a more passive study and analysis role to one that actively seeks to convey reliable and impartial information to interested parties in a way that is responsive to social, political, and economic needs. There are numerous ways to define the agency's outreach activities that would meet this broad objective, but a simple way to characterize the committee's recommendation as to what the outreach role of the USGS should be is that outreach efforts should help decision makers and the general population avoid being surprised.

The contrasts between the two examples of successful and responsive outreach programs reviewed in the Louisiana HydroWatch program and the Cascade Volcano Outreach Project illustrate the essential elements common to effective outreach. In both regions, communities were faced with real and potential hydrologic hazards, posing significant threats to life and property. In both cases, USGS was in the position to target its expertise and effectively allocate resources to provide the reliable impartial information crucial for public education, regional planning, and emergency response. The outreach products developed in each study represented a thoughtful balance between the needs of the outreach audience, the unique expertise of the Survey, and the inevitable competition for limited resources (both within USGS and among cooperating agencies and stakeholders). The Survey's scientific and technical expertise is unique in the Cascades

Volcano example, resulting in a vigorous and sustained outreach effort by USGS. In contrast, the Louisiana HydroWatch is a prime example illustrating the value and opportunities to substantially enhance outreach and hazard communication by closely coordinating USGS expertise with the complementary strengths and capabilities of both federal resource agencies like the National Weather Service and the Corps of Engineers, and local and regional resource management and emergency response agencies.

These two outreach programs are not special cases: rather they are examples of effective outreach efforts, well matched to the audience, information needs, and resources, that have been effected through nontraditional activities. Both examples illustrate outreach programs that (1) identified the audiences for hazard outreach and their respective needs; (2) targeted unique scientific and technical expertise within USGS to those needs; and (3) effectively balanced the competing demands for outreach with the limited resources and complementary capability of both federal and nonfederal cooperating agencies.

Overall, successful outreach in the area of hydrologic hazards requires methods to help decision makers better visualize the probabilities and consequences of hazards locally. The USGS can take the lead in improving hydrologic hazard understanding through improved visualization approaches that integrate expertise existing in the various divisions of the agency in long-term monitoring, mapping, and process modeling. The flood tracking charts are a good example of how the general community can use real-time data and flood predictions to make decisions regarding safety. In addition to the Internet, real-time data need to be disseminated along other available avenues. There is a critical need for improved methods of risk communication. Most people do not understand simple concepts of probability. Consequently, there is a general misconception as to what the 100-year floodplain designates and concepts of drought are even more difficult. It would be highly useful to explain flood risk by creating various scenarios that explain consequences in terms of a given set of conditions or "what ifs." These scenarios might be very effectively illustrated by graphical computer simulations or animations. As the USGS explores approaches to better communicate hydrologic risk, a useful reference to consult would be the 1989 NRC report *Improving Risk Communication* (NRC, 1989).

While outreach is an important new direction for the USGS, the agency should understand that outreach efforts must be a logical and comfortable extension of its strengths in data collection and interpretation. If the USGS views outreach as a completely new agency mission, it will detract resources from its core programs. Outreach should require marginal adjustments in programs and budgets, consistent with the core science programs of the agency. In developing its outreach program the USGS must identify and target the audiences it wishes to serve and define outreach products that draw from existing strengths of the agency. Throughout this process the USGS should be assessing and refining the quality and utility of its outreach programs.

6

Conclusions and Recommendations

The U.S. Geological Survey's (USGS) programs in hydrologic hazards are necessarily interdisciplinary and offer many opportunities for collaborative work between personnel of the Water Resources Division, Geologic Division, Biological Resources Division, and National Mapping Division, as well as with outside cooperators and colleagues at the National Weather Service, the U. S. Army Corps of Engineers, state agencies, and universities. Among these and other institutions, the USGS is unique in its motivation and capability to explore and understand the physical processes involved in hydrologic hazards resulting from extreme hydrologic events. As such, the USGS has a critical role in collecting data relevant to hydrologic hazards and translating those data into information useful to other public and private institutions and to the public at large.

The USGS role in hydrologic hazards has multiple products and multiple clients (see Figure 6.1). As the nation's principal collector of water resources data, the USGS has a long tradition and continuing responsibility for collecting field data that are relevant to hydrologic hazards. Once collected, data such as water elevations and streamflows should be made immediately available to interested parties for use in decision making. The data are also critical to research dealing with predicting and understanding causes and consequences of hydrologic hazards. Interpreted in the context of other data and existing models, each new dataset adds new information about hydrologic hazards. The information developed though data interpretation, analysis, and interaction with user groups is also critical to the design of improved data collection programs.

Information generated through hydrologic hazards programs will have mul-

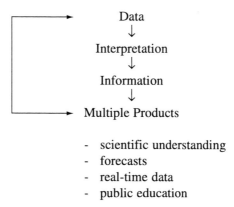

FIGURE 6.1 The USGS roles in hydrologic hazards.

tiple users and multiple products. Examples of products and users of this information include the following:

• *Federal, state, and local emergency management agencies (FEMA, National Weather Service).* Such agencies require reliable real-time datasets and statistical distributions (current water levels, stream rating curves) for effective forecasting and analysis of the consequences of extreme events. Such users are most often interested in basic data with little or no interpretation or analysis.

• *Federal, state, and local water resource management agencies (U.S. Army Corps of Engineers, U.S. Environmental Protection Agency, state and local environmental institutions), industries, and technical consulting firms.* These institutions often require longer-term information about environmental conditions (low-flow duration curves, historical low- and high-water levels, water quality information) for use in developing and implementing environmental regulations and in the planning, design, and operation of water-related facilities. Often these clients require technical summaries and interpretations of data rather than the actual raw data.

• *General public, private citizens, educators, law enforcement officials, the news media, nonprofit environmental groups, private insurers, etc.* These parties rely on the USGS to provide not only raw data but also interpretation and explanation of the meaning of the data, which then becomes the basis for planning, investment, and management of programs related to hydrologic hazards. These users often require explanations of the basic principles of hydrology, probability, and flood/drought consequences. What is needed are ways to help decision makers and citizens better visualize probabilities and consequences of hazards within the context of their particular situations.

• *The scientific community.* The larger scientific community interested in hydrologic processes requires the data, interpretations, and models developed by USGS scientists to advance the state of the science and develop a better understanding of the physical causes and consequences of extreme hydrologic events. USGS scientists should continue to publish scientific reports and participate in scientific conferences to share findings with their colleagues.

FLOODS AND DROUGHTS

Role of the USGS

Through its national streamflow measurement program, the USGS should continue to be the nation's primary supplier of reliable streamflow and water-level data. This information will continue to be used by the National Weather Service and other agencies for flood forecasting, design purposes, and for forecasting low flows and water levels in times of drought.

The committee believes the USGS has an additional important role in the measurement, documentation, and analysis of extreme hydrologic events, both during and after they occur. The USGS is ideally positioned to collect and archive the critical hydrologic information necessary to improve our understanding of how and why such extreme events happen and to improve our ability to predict these events. In particular, the USGS should initiate process studies on high-magnitude, low-frequency events (both floods and droughts) to improve our overall technical understanding of how and why these events occur.

Recommendations

USGS flood and drought science should focus on the following areas:

• *Maintaining the integrity and continuity of the national stream gaging network.* The USGS stream gaging network is a unique and irreplaceable source of primary data supporting planning, research, and management for hydrologic hazards. It is of critical national importance that this source of consistent and reliable hydrologic data be maintained, both as the foundation for other hydrologic activity conducted by USGS and as the basis for planning and operations carried out by countless other public and private entities.

• *Improved stream gaging network design, measurement techniques, and instrumentation for the measurement of streamflow and stream stage.* Obtaining streamflow data, particularly if the current climate of funding cutbacks continues, will require new and innovative instrumentation and monitoring techniques. Recent advances in microprocessors, electronics, and satellite communication are leading to a new generation of reliable, precise, and relatively inexpensive field instrumentation equipment. Research in gaging network optimization will help obtain the best information for a given level of investment.

• *Postaudits of technical response and prediction of major floods.* Postaudits of major floods such as the 1997 Red River flood in North Dakota should be conducted, with the goal of improving data collection, flood forecasting, and risk communication. If there were problems with the flood forecasts, why did they occur and how could they have been avoided? Were appropriate data collected before, during, and after the flood to improve our understanding of the causes and effects of the flood?

• *Improved discharge measurements of extreme floods.* Currently used methods of flood forecasting usually employ the traditional stage-discharge relations developed at stream gages. The stage-discharge relationship often breaks down during extreme events that are beyond the range of normally measured stages. Techniques such as acoustic doppler profiling or process modeling may allow rating curves to be extended beyond their traditional limits.

• *Improved approaches for regional flood-frequency estimation.* Opportunities to improve Bulletin 17-B (USIACWD, 1982), better estimation of flood-frequency relations in ungaged watersheds, and better ways to incorporate historical and paleoflood data into flood-frequency analyses need to be pursued.

• *Improved methods for drought forecasting.* Emphasis should be placed in integrative studies linking climate variability, terrestrial hydrology and surface-ground water interactions at the regional scale.

• *Investigations of long-term stationarity of floods and droughts.* Current frequency estimation methods usually assume that flood peak heights, discharges, and recurrence intervals follow statistical distributions that do not change with time. This assumption of stationarity is not consistent with the natural variability exhibited by hydrologic data when long records are available. How are flood heights responding to short-term and long-term changes in land use, urbanization, weather patterns, and water management regimes? What is a reasonable period of record with which to evaluate flood risks? Such investigations could also consider prehistoric information such as that offered by tree ring data and paleo lake studies to estimate the magnitude and frequency of very rare flows.

• *Improved techniques for low-flow frequency analysis, and its relevance to instream flow management and ecologically based regulatory criteria.* Low-flow frequency analysis, commonly draws on flood-frequency techniques developed to estimate long-term averaged exceedance probabilities or drought risks. The Water Resources Division and the Biological Resources Division should collaborate to improve our understanding of the relationships between simple quantile estimates of low-flow frequencies and the response and sensitivity of aquatic ecosystems to the magnitude, duration, and interarrival times of low-flow regimes.

ASSISTANCE IN RISK-BASED DECISION MAKING

Role of the USGS

As the nation's principal provider of geologic and hydrologic data, the USGS should position itself as a respected national resource for information needed for risk-based decision making with respect to hydrologic hazards. To be such a resource, the agency will need to continue to maintain a staff of respected unbiased scientists working in the hydrologic hazards area. In addition, the agency will need to support integrated database management systems to inventory, store, and make accessible regularly collected meteorological and hydrologic information, on a watershed-by-watershed basis, with easy linkages between weather, topographic, streamflow, and reservoir management data, probably through the use of geographic information systems (GISs). This information should be readily accessible to researchers involved in the development and testing of statistical and deterministic models for hydrologic processes.

Recommendations

• The USGS should conduct research on techniques for estimating the probability and magnitude of extreme hydrologic events in the context of risk-based decision making.
• The USGS should evaluate how long-term cumulative impacts of land use, river regulation, or climate change can affect and contribute to hydrologic hazards.
• The USGS should work to improve integrated risk and process models related to floods and droughts.

WATER RESOURCES DATA

Role of the USGS

The USGS has long and broad experience in collecting, managing, and disseminating water resources data. These data functions continue to be a critical part of the hydrologic hazards program and must be maintained. The recent integration of the Biological Resources Division into the USGS provides a new opportunity to improve the integration of water resources data with biological and ecological data.

Recommendations

• The USGS should place new emphasis on rapid data acquisition and retrieval during extreme events. Real-time dissemination of provisional data acquired during floods will permit better decision making.

• The USGS should explore methods for integrating datasets over several scientific disciplines. Understanding hydrologic hazards usually requires data from several different scientific areas. For example, landslide forecasting might require information collected by hydrologists, geologists, engineers, soil scientists, and climatologists. Historically, such data have been stored in different formats by different agencies. GIS technology may offer techniques for integrating, analyzing, and displaying such dissimilar datasets for improved analysis of hydrologic hazards.

PUBLIC OUTREACH AND HAZARDS COMMUNICATION

Role of the USGS

The ultimate goal of the hydrologic hazards program is to assist in protecting the lives and property of citizens from naturally occurring hazards while maintaining and protecting ecological communities. This requires that hazards information and research results be communicated to the public, and to public officials, in a timely and understandable manner. The translation of research science to public understanding and public policy is often difficult and usually requires special skills. The revolution in information technologies has stimulated dramatic changes in how the USGS collects and disseminates its data and in the audiences interested in this information. In response, the emphasis of the USGS mission to provide the nation with reliable and impartial information to reduce loss of life and property and to preserve resources has shifted from a more passive role of study and analysis to one that actively seeks to convey information in a way that is responsive to the social, political, and economic needs of particular communities. It is critical that the USGS maintain and develop liaisons with outreach specialists in other federal agencies, federal and state extension programs, universities, state and local units of government, commercial organizations, and citizen interest groups. Individual USGS scientists should develop networks of public contacts and should be encouraged to participate in public discussions of hazards issues.

Recommendations

The USGS should continue to employ both new and existing technologies to enhance the communication of hazards information to multiple and diverse client groups. Examples of communications efforts include the following:

• Direct involvement of USGS scientists in communication with the public.
• Development of hazards maps for various parts of the United States. The format, scale, locations, and types of hazards shown on such maps may vary from place to place.

• Integration of hazards maps with electronic databases and GISs. Examples from the Cascades Volcano Observatory and Louisiana HydroWatch program illustrate the utility of hazard mapping oriented for the public. These outreach products represent an intelligent match for the mapping, data collection, and analytical strengths of the USGS. The agency should expand this effort into several other areas. What is the best way to communicate hazards information from agency to agency? How can hazards information best be integrated and utilized in public policy decision making? The USGS mission in outreach should not be to direct particular decisions about hydrologic hazard management. Instead, the agency should ensure that decisions are made with the best possible understanding of probabilities and magnitudes of adverse consequences. To put it another way, outreach should help decision makers avoid being "surprised."

• The USGS is progressive in its use of the Internet for the presentation of real-time and historical data. These programs are impressive, and the agency should continue to expand these efforts as well as add interpretations to its hydrologic data. In addition to the Internet, real-time data need to be disseminated along other available avenues.

• There is a critical need for improved methods of risk communication. Since most people do not understand the simple concepts of probability (e.g., 100-year floodplain), it would be highly useful to explain flood risk in terms of possible or probable consequences by creating various scenarios. These scenarios might then be very effectively illustrated by graphical computer simulations or animations. What is needed are ways to help decision makers better visualize the probabilities and consequences of hydrologic hazards. The USGS can take the lead in improving this understanding by developing effective "visualization" approaches.

References

Alley, W. M. 1984. The Palmer Drought Index: Limitations and assumptions. J. Climate Appl. Meteor. 23:1100-1109.

Baker, V. R. 1998. Paleohydrology and the hydrological sciences. Pp. 1-10 in Paleohydrology and Environmental Change, G. Benito, V. R. Baker, and K. J. Gregory, eds. Chichester, England: Wiley and Sons.

Barros, A. P., and R. J. Kuligowski. 1998. Orographic effects during a severe wintertime storm in the Appalachian Mountains. Monthly Weather Review 126:2648-2672.

Barros, A. P., and D. P. Lettenmaier. 1994. Dynamic modeling of orographically-induced precipitation. Rev. Geophysics 32:265-284.

Booy, C., and D. R. Morgan. 1985. The effect of clustering of flood peaks on a flood risk analysis for the Red River. Can. J. Civil Eng. 12:150-165.

Bosch, J. M., and J. D. Hewlett. 1982. A review of catchment experiments to determine the effects of vegetation changes on water yield and evapotranspiration. J. Hydrol. 55:3-23.

Bradley, A. A. 1998. Regional frequency analysis methods for evaluating changes in hydrologic extremes. Water Resour. Res. 34(4):741-750.

Burton, I., R. W. Kates, and G. F. White. 1993. The Environment as Hazard, Second Edition. New York: Guilford Press.

Canadian Electricity Association (CEA). 1994. Circulation Patterns for Streamflow and Temperature Predictions in Canadian Surface Climate. R&D Report 9206 g 931. CEA, Montréal, Québec.

Cayan, D. R., and Webb, R. H. 1992. El Niño/Southern Oscillation and streamflow in the western United States. Pp. 29-68 in El Niño, Historical and Paleoclimatic Aspects of the Southern Oscillation, H. F. Diaz and V. Markgraf, eds. Cambridge, England: Cambridge University Press.

Changnon, S. A. 1989. Midwestern drought conditions–1988. Pp. 17-40 in Drought and Climate Change: Miscellaneous Papers on the 1988 Drought and the Issue of Future Climate Change by the Staff Members of the Midwestern Climate Change Center. Research Report 89-02. Champaign, Illinois.

Cleave, M. K., and D. W. Stahle. 1989. Tree ring analysis of surplus and deficit runoff in the White River, Arkansas. Water Resour. Res. 25:1391-1401.

Clement, R. W. 1987. Floods in Kansas and Techniques for Estimating Their Magnitude and Frequency. U.S. Geological Survey Water Resources Investigations Report 87-4008.

Code of Federal Regulations (CFR). 1992. U.S. Annotated Code of Federal Regulations, 57FR60848. P. 60916. Dec. 22. Washington, D.C.

Cohn, T. A., and J. R. Stedinger. 1987. Use of historical information in a maximum likelihood framework. J. Hydrol. 96(1-4):215-223.

Collier, M., R. H. Webb, and J. C. Schmidt. 1996. Dams and Rivers, A Primer on the Downstream Effects of Dams. U.S. Geological Survey Circular 1126. Reston, Virginia.

Costa, J. E. 1986. A history of paleoflood hydrology in the United States, 1800-1970. EOS 67:425-430.

Curtis, G. W. 1987. Techniques for estimating flood-peak discharges and frequencies on rural streams in Illinois. U.S. Geological Survey Water-Resources Investigations Report 87-4207. Reston, Virginia.

Dalrymple, T., and M. A. Benson. 1967. Measurement of peak discharges by the slope-area method. U.S. Geological Survey Techniques of Water-Resources Investigations, Book 3, Chap. A2. Reston, Virginia.

Dracup, J. A., and E. Kahya. 1994. The relationships between U.S. streamflow and the La Niña events. Water Resour. Res. 30A(7):2133-2141.

Dracup, J. A., K. S. Lee, and E. G. Paulson, Jr. 1980. On the definition of droughts. Water Resour. Res. 16(2):297-302.

Dyson, L. K. 1988. History of Federal Drought Relief Programs, Report No. AGES880914. U.S. Department of Agriculture, Washington, D.C.

Ely, L. L. 1997. Response of extreme floods in the southwestern United States to climatic variations in the late Holocene. Geomorphology 19:175-201.

Ely, L. L., Y. Enzel, V. R. Baker, and D. Cayan. 1993. A 5000-year record of extreme floods and climate change in the southwestern United States. Science 262:410-412.

Ely, L. L., Y. Enzel, and D. Cayan. 1994. Anomalous North Pacific circulation and large winter floods in the southwestern United States. J. Climate 7:977-987.

Ely, L. L., Y. Enzel, V. R. Baker, V. S. Kale, and S. Mishra. 1996. Paleoflood evidence of changes in the magnitude and frequency of monsoon floods on the Narmada River, central India. Geo. Soc. Am. Bull. 108:1134-1148.

Enzel, Y., L. L. Ely, P. K. House, V. R. Baker, and R. H. Webb. 1993. Paleoflood evidence for a natural upper bound to flood magnitudes in the Colorado River Basin. Water Resour. Res. 29:2287-2297.

Enzel, Y., L. L. Ely, P. K. House, and V. R. Baker. 1996. Magnitude and frequency of Holocene paleofloods in the southwestern United States: A review and discussion of implications. Pp. 121-137 in Global Continental Changes: The Context of Paleohydrology, J. Branson, A. G. Brown, and K. J. Gregory, editors. Geological Society of London Special Publ. 115.

Federal Interagency Floodplain Management Task Force (FIFMTF). 1992. Floodplain Management in the United States: An Assessment Report. Washington, D.C.

Ferguson, B. K., and P. W. Suckling. 1991. Changing rainfall-runoff relationships in the urbanizing Peachtree Creek watershed, Atlanta, Georgia. Water Resour. Bull. 26(2):313-322.

Fulford, J. M. 1994. User's Guide to SAC, A Computer Program for Computing Discharge by the Slope-Area Method. U.S. Geological Survey Open-File Report 94-360. Reston, Virginia.

Godschalk, D. R., T. Beatley, P. Berke, D. J. Brower, and E. J. Kaiser. 1997. Making Mitigation Work: Recasting Natural Hazards Planning and Implementation. Final report to the National Science Foundation, Center for Urban and Regional Studies, University of North Carolina, Chapel Hill.

Graf, J. B., R. H. Webb, and R. Hereford. 1991. Relation of sediment load and flood-plain forma-
tion to climatic variability, Paria River drainage basin, Utah and Arizona. Geol. Soc. Am. Bull.
103:1405-1415.

Grigg, N. S. 1993. Drought and water-supply management: Roles and responsibilities. J. Water
Resour. Plan. Mgmt. 119(5):531-541.

Grigg, N. S. 1996. Water Supply Management: Principles, Regulations, and Cases. New York:
McGraw-Hill.

Grimm, M. M., E. E. Wohl, and R. D. Jarrett. 1995. Coarse-sediment distribution as evidence of an
elevation limit for flash flooding, Bear Creek, Colorado. Geomorphology 14(12):199-210.

Guimaraes, W. B., and L. R. Bohman. 1991. Techniques for estimating magnitude and frequency
of floods in South Carolina, 1988. U.S. Geological Survey Water-Resources Investigations
Report 91-4157. Reston, Virginia.

Hirsch, R. M. 1978. Risk analyses for a water-supply system—Occoquan Reservoir, Fairfax and
Prince William counties, Virginia. Hydrol. Sci. 23(4):475-505.

Hirschfield, D. M., and W. T. Wilson. 1960. A comparison of extreme rainfall depths from tropical
and nontropical storms. J. Geophy. Res. 65:959-982.

Holnbeck, S. R., and C. Parrett. 1997. Method for rapid estimation of scour at highway bridges
based on limited site data. U.S. Geological Survey Water Resources Investigations Report 96-
4310. Reston, Virginia.

Hrezo, M. S., P. G. Bridgeman, and W. R. Walker. 1986. Managing droughts through trigger
mechanisms. J. Am. Water Works Assoc. (June):46-51.

Interagency Floodplain Management Review Committee (IFMRC). 1994. Sharing the Challenge:
Floodplain Management into the 21st Century. Washington, D.C.:U. S. Government Printing
Office.

Jarrett, R. D. 1990. Paleohydrology used to define the spatial occurrence of floods. Geomorphology
3(2):181-195.

Jarrett, R. D. 1991. Paleohydrology and its value in analyzing floods and droughts. U.S. Geological
Survey Water Supply Paper 2375. Pp. 105-116. Reston, Virginia.

Jarrett, R. D. 1993. Flood elevation limits in the Rocky Mountains. Pp. 180-185 in Engineering
Hydrology, C. Y. Kuo, ed. Proceedings of a symposium sponsored by the Hydraulics Division
of the American Society of Civil Engineers, San Francisco, July 25-30. New York: American
Society of Civil Engineers.

Jennings, M. E., W. O. Thomas, and H. C. Riggs. 1994. Nationwide Summary of U.S. Geological
Survey Regional Regression Equations for Estimating Magnitude and Frequency of Floods for
Ungaged Sites, 1993. U.S. Geological Survey Water-Resources Investigations Report 94-
4002. Reston, Virginia.

Karl, T. R. 1983. Some spatial characteristics of drought duration in the United States. J. Climate
and Appl. Meteorol. 22:1356-1366.

Kirby, W. H. 1981. Annual flood frequency analysis using U.S. Water Resources Council guidelines (program J407). U.S. Geological Survey Open-File Report 79-1336-I, WATSTORE
User's Guide, vol. 4, Chap. I. Reston, Virginia.

Klemes, V. 1986. Dilettanism in hydrology: Transition or destiny? Water Resour. Res. 22(9):
177S-188S.

Kunkel, K. E., and J. R. Angel. 1989. Perspective on the 1988 midwestern drought. EOS. Sept. 5.

Kunkel, K. E., S. A. Chagnon, and P. J. Lamb. 1989. The 1988 midwestern drought: Key elements
and impacts. Paper presented at the Fifteenth Annual Illinois Crop Protection Workshop,
Cooperative Extension Service. Illinois Natural History Survey, University of Illinois at Ur-
bana-Champaign.

Landers, M. N., and K. V. Wilson, Jr. 1991. Flood characteristics of Mississippi streams. U.S.
Geological Survey Water-Resources Investigations Report 91-4037. Reston, Virginia.

Landwehr, J. M., and J. R. Slack. 1990. Evidence of climate change or climate variation in discharge records from the United States. Paper presented at the Chapman Conference on Hydrologic Aspects of Global Climate Change, Lake Chelan, Wash. June 12-14 American Geophysical Union.

Liu, S., and J. R. Stedinger. 1991. Low flow frequency analysis with ordinary and tobit regression. Pp. 27-31 in Water Resources Planning and Management and Urban Water Resources, Proceedings of the 18th Annual Conference and Symposium, J. L. Anderson, ed. New York: American Society of Civil Engineers.

Lott, N. 1993. The Summer of 1993: Flooding in the Midwest and Drought in the Southeast. TR 93-04 National Oceanographic and Atmospheric Administration National Climate Data Center. Washington, D.C.

Ludwig, A. H., and G. E. Tasker. 1993. Regionalization of low flow characteristics of Arkansas streams. U.S. Geological Survey Water-Resources Investigations Report 93-4013. Reston, Virginia.

Matalas, N. C. 1991. Drought description. Stoch. Hydrol. Hydraul 5:255-260.

McNab, A. L. 1989. Climate and drought. EOS 70(40):873, 882-883.

Medina, K. D., and G. D. Tasker. 1985. Analysis of Surface Water Data Network in Kansas for Effectiveness in Providing Regional Information. U.S. Geological Survey Water Supply Paper 2203. Reston, Virginia.

Moody, J. A., and R. H. Meade. 1990. Channel changes at cross sections of the Powder River between Moorhead and Broadus, Montana, 1975-88. U.S. Geological Survey Open-File Report 89-407. Reston, Virginia.

Moss, M. E., and G. D. Tasker. 1991. An intercomparison of hydrological network-design technologies. Hydrol. Sci. J. 36:201-213.

Moss, M. E., and G. D. Tasker. 1995. HYNET—An intercomparison of hydrologic network-design technologies in World Meteorological Organization. Tech. Rep. Hydrol. Water Resour. 50:17.1-17.5.

Namias, J. 1985. Factors responsible for droughts. Pp. 27-64 in Hydrologic Aspects of Drought, M. A. Beran and J. A. Rodier, eds. United National Educational, Scientific, and Cultural Organization. Paris, France.

National Drought Mitigation Center (NDMC). 1995. Understanding and defining drought. http://enso.unl.edu/ndmc/enigma/def2.htm.

National Drought Mitigation Center (NDMC). 1996. Drought indices. http://enso.unl.edu/ndmc/enigma/indices.htm.

National Drought Mitigation Center (NDMC). 1997. Understanding ENSO and forecasting drought. http://enso.unl.edu/ndmc/enigma/elnino.htm.

National Research Council. 1988. Estimating Probabilities of Extreme Floods: Methods and Recommended Research. Washington, D.C.: National Academy Press.

National Research Council. 1989. Improving Risk Communication. Washington, D.C.: National Academy Press.

National Research Council. 1991. Opportunities in the Hydrologic Sciences. Washington, D.C.: National Academy Press.

National Research Council. 1992. Regional Hydrology and the USGS Stream Gaging Network. Washington, D.C.: National Academy Press.

National Research Council. 1995. Flood Risk Management and the American River Basin: An Evaluation. Washington, D.C.: National Academy Press.

National Weather Service (NWS). 1994. The Great Flood of 1993. Washington, D.C.:NWS.

O'Connor, J. E., L. L. Ely, E. E. Wohl, L. E. Stevens, T. S. Melis, V. S. Kale, and V. R. Baker. 1994. 4500-year record of large floods on the Colorado River in the Grand Canyon, Arizona. J. Geology 102:1-9.

O'Grady, K., and L. Shabman. 1994. Uncertainty and time preference in shore protection. Pp. 136-154 in Risk Based Decision Making in Water Resources, Y. Haimes et al., eds. New York: American Society of Civil Engineers.

Pizzuto, J. E. 1994. Channel adjustments to changing discharges, Powder River, Montana: Geol. Soc. Am. Bull. 106:1494-1501.

Riebsame, W. E., S. A. Changnon, Jr., and T. R. Karl. 1991. Drought and Natural Resources Management in the United States. Boulder, Colo.: Westview Press.

Ruddy, B. C., and K. J. Hitt. 1990. Summary of Selected Characteristics of Large Reservoirs in the United States and Puerto Rico, 1988. Open File Report No. 90-163. Denver, Colo.: U.S. Geological Survey.

Salas, J. D., E. E. Wohl, and R. D. Jarrett. 1994. Determination of flood characteristics using systematic, historical, and paleoflood data. Pp. 111-134 in Coping with Floods. Dordrecht: Kluwer Academic Publishers.

Schoemaker, P. J. H., and H. C. Kunruther. 1979. An experimental study of insurance decisions. J. Risk Ins. 46:608-618.

Seaburn, G. E. 1969. Effects of Urban Development on Direct Runoff to East Meadow Brook. U.S. Geological Survey Professional paper 627-B. Reston, Virginia

Slack, J. R., and J. M. Landwehr. 1992. Hydro-Climatic Data Network (HCDN): A U.S. Geological Survey Streamflow Data Set for the United States for the Study of Climate Variations, 1874-1988. U.S. Geological Survey Open-File Report 92-129. Reston, Virginia.

Slovic, P. 1977. Preference for insuring against probable small losses: Insurance and implications. J. Risk In. 44:237-258.

Stedinger, J. R. 1996. Expected probability and annual damage estimators. J. Water Resour. Plan. Mgmt. Pp. 125-135.

Stedinger, J. R., and V. R. Baker. 1987. Surface water hydrology: Historical and paleoflood information. Rev. Geophysics 25:119-124.

Stedinger, J. R., and G. D. Tasker. 1985. Regional hydrologic analysis 1. Water Resour. Res. 21(9):1421-1432.

Stedinger, J. R., and G. D. Tasker. 1986. Regional hydrologic analysis 2. Water Resour. Res. 22(10):1487-1499.

Stockton, C. W., W. R. Boggess, and D. M. Meko. 1985. Climate and Tree Rings in Paleoclimate Analysis and Modeling, A. D. Hecht, ed. New York: Wiley.

Tasker, G. D. 1991a. Evaluating drought risks for large highly regulated basins using monthly water-balance modeling. Pp. 317-322 in Proceedings of the U.S.-China Bilateral Symposium on Droughts and Semi-Arid Region Hydrology. U.S. Geological Survey Open-File Report 91-244. Reston, Virginia.

Tasker, G. D. 1991b. Identifying stream gages to operate for regional information. Bridge Hydrol. Res. 1319:131-136.

Tasker, G. D. 1993. Some hydrologic effects of climate change for the Appalachicola, Chattahoochee, and Flint River basins. Pp. 61-61 in Georgia Water Resources Conference, Athens, Georgia, K. J. Hatcher, ed. University of Georgia, Institute of Natural Resources.

Tasker, G. D., and R. M. Slade, Jr. 1994. An interactive regional regression approach to estimating flood quantiles. Pp. 782-785 in Water Policy and Management: Solving the Problems, D. G. Gontane and H. N. Tuvel, eds. 21st Annual Conference Proceedings. American Society of Civil Engineers.

Tasker, G. D., and J. R. Stedinger. 1989. An operational GLS model for hydrologic regression. J. Hydrol. 111:361-375.

Tasker, G. D., and J. R. Stedinger. 1992. Generalized least squares analyses for hydrologic regionalization. Pp. 7-12 in Hydraulic Engineering: Saving a Threatened Resource in Search of Solutions, M. Jennings and N. G. Bhowmik, eds. Proceedings of the Hydraulic Engineering Sessions at Water Forum 1992, Baltimore.

Tasker, G. D., M. Ayers, D. Wolock, and G. McCabe. 1991. Sensitivity of drought risks in the Delaware River basin to climate change. Pp. 153-159 in Technical and Business Exhibition and Symposium. Huntsville, Alabama: Huntsville Association of Technical Societies.

Tasker, G. D., S. A. Hodge, and C. S. Barks. 1996. Region of influence regression for estimating the 50-year flood at ungaged sites. Water Resour. Bull. 32(1):163-170.

U.S. Army Corps of Engineers. 1994. National Study of Water Management During Drought—Report to Congress. Report No. IWR 94-NDS-12. Fort Belvoir, Va.: Institute for Water Resources.

U.S. Environmental Protection Agency (USEPA). 1990. DFLOW User's Manual. EPA 600/8-90/051. Washington, D.C.: USEPA.

U.S. Environmental Protection Agency (USEPA). 1991. Technical Support Document for Water Quality-Based Toxics Control. EPA/505/2-90-001. Washington, D.C.: USEPA.

U.S. Geological Survey (USGS). 1996. Strategic Plan for the U.S. Geological Survey: 1996 to 2005. Reston, Va.: USGS.

U.S. Interagency Advisory Committee on Water Data (USIACWD). 1982. Guidelines for Determining Flood Flow Frequency. Bulletin 17-B of the Hydrology Subcommittee. Reston, Va: U.S. Geological Survey, Office of Water Data Coordination.

Vogel, R. M., and C. N. Kroll. 1990. Generalized low-flow frequency relationships for ungaged sites in Massachusetts. Water Resour. Bull. 26(2):241-253.

Wahl, K. L., W. O. Thomas, Jr., and R. M. Hirsch. 1995. The Stream-Gaging Program of the U.S. Geological Survey. U.S. Geological Survey Circular 1123. Reston, Virginia.

Wallace, J. R. 1971. The Effects of Land Use Changes on the Hydrology of an Urban Watershed. OWRR Report Project C-1786. Atlanta, Georgia: School of Civil Engineering, Georgia Institute of Technology.

Waylen, P., and M. K. Woo. 1982. Prediction of annual floods generated by mixed processes. Water Resour. Res. 18(4):1283-1286.

Webb, R. H., and J. L. Betancourt. 1990. Climatic effects on flood frequency: An example from southern Arizona. Pp. 61-66 in Proceedings of the Sixth Annual Pacific Climate (PACLIM) Workshop, Asilomar, California, March 5-8, 1989, J. L. Betancourt and A. M. MacKay, eds. California Department of Water Resources, Interagency Ecological Studies Program, Technical Report 23.

Webb, R. H., and J. L. Betancourt. 1992. Climatic Variability and Flood Frequency of the Santa Cruz River, Pima County, Arizona. U.S. Geological Survey Water-Supply Paper 2379. Reston, Virginia.

Wilhite, D. A. 1993. Drought Assessment, Management, and Planning: Theory and Case Studies. Boston: Kluwer Academic Publishers.

Wolock, D. M., G. J. McCabe, G. D. Tasker, and M. E. Moss. 1993. Effects of climate change on water resources in the Delaware River Basin. Water Resour. Bull. 29(3):475-486.

Biographical Sketches of
Committee Members

KENNETH R. BRADBURY (*Chairman*) is a research hydrogeologist/professor with the Wisconsin Geological and Natural History Survey, University of Wisconsin-Extension, in Madison. He received his Ph.D. (hydrogeology, 1982) from the University of Wisconsin-Madison, his A.M. (geology, 1977) from Indiana University, and his B.A. (geology, 1974) from Ohio Wesleyan University. His current research interests include ground water flow in fractured media, ground water recharge processes, wellhead protection, and the hydrogeology of glacial deposits.

VICTOR R. BAKER is regents professor and head of the Department of Hydrology and Water Resources at the University of Arizona. He is also professor of geosciences and professor of planetary sciences at the University of Arizona. His research interests include geomorphology, flood geomorphology, paleohydrology, Quaternary geology, natural hazards, geology of Mars and Venus, and philosophy of earth and planetary sciences. He has spent time as a geophysicist for U.S. Geological Survey and as an urban geologist. He has served on various committees and panels of the National Research Council, including the Panel on Alluvial Fan Flooding, the Panel on Global Surficial Geofluxes, and the Panel on Scientific Responsibility and Conduct of Research. He formerly chaired the U.S. National Committee for the International Union for Quaternary Research (INQUA) and served on the Global Change Committee Working Group on Solid Earth Processes. Dr. Baker is currently president of the Geological Society of America and president of the INQUA Commission on Global Continental

Paleohydrology. He holds a B.S. from Rensselaer Polytechnic Institute and a Ph.D. from the University of Colorado.

ANA P. BARROS is an associate professor of civil engineering at the Pennsylvania State University. She received a diploma in civil engineering from the University of Porto (Portugal) in 1985, an M.S. in hydraulics/ocean engineering from the University of Porto in 1988, an M.S. in environmental science and engineering from Oregon Graduate Institute in 1990, and a Ph.D. in civil engineering from the University of Washington in 1993. Dr. Barros's research interests are environmental fluid mechanics, land-atmosphere interactions, macroscale hydrology, hydrometeorology of mountainous regions, hydrologic extremes (floods and droughts), climate variability, and remote sensing.

MICHAEL E. CAMPANA is director of the Water Resources Program and professor of earth and planetary sciences at the University of New Mexico. His current interests are hydrologic system-aquatic ecosystem interactions, regional hydrogeology, environmental isotope hydrology, and the hydrology of arid and tropical regions. He teaches courses in water resources management, hydrogeology, subsurface fate and transport processes, environmental mechanics, and geological fluid mechanics. He was a Fulbright scholar to Belize in 1996. Dr. Campana received a B.S. in 1970 in geology from the College of William and Mary, an M.S. in hydrology in 1973, and a Ph.D.in hydrology in 1975 from the University of Arizona.

KIMBERLY A. GRAY is an associate professor of environmental engineering in the Department of Civil Engineering at Northwestern University. She received her Ph.D. from Johns Hopkins in 1988, an M.S. from the University of Miami in 1983 in civil engineering, and her B.A. in 1978 in biology from Northwestern University. Dr. Gray teaches physicochemical processes, aquatic chemistry, environmental analytical chemistry, and drinking water treatment design. Her research entails experimental study of both engineered and natural processes. She studies the characteristics of natural organic matter in surface waters, wetlands, and treatment systems by pyrolysis-GC-MS. Other topics of her research include the use of semiconductors to photocatalyze the destruction of hazardous chemicals, the application of ionizing radiation to reductively dechlorinate pollutants in soil matrices, and the ecotoxicology of PCBs in periphytic biolayers.

C. THOMAS HAAN is the regents professor and Sarkeys distinguished professor in the Department of Biosystems and Agricultural Engineering at Oklahoma State University. He received his Ph.D. in agricultural engineering from Iowa State University in 1967. Dr. Haan's research interests are hydrology, hydrologic and water quality modeling, stochastic hydrology, and risk analysis.

He has served as a consultant to several national and international agencies. Dr. Haan is a member of the National Academy of Engineering.

DAVID H. MOREAU is professor in the Departments of City and Regional Planning and Environmental Sciences and Engineering at the University of North Carolina, Chapel Hill. Chair of the Department of City and Regional Planning, Dr. Moreau received a B.Sc. (civil engineering, 1960) from Mississippi State University, an M.Sc. (civil engineering, 1963) from North Carolina State University, an M.Sc. (engineering, 1964) from Harvard University, and a Ph.D. (water resources, 1967) from Harvard University. Dr. Moreau has been a consultant to the United Nations Development Program, Water Management Models for Water Supply; New York City, review of water demand projections; and Water for Sanitation and Health Program (AID), financing of water supply and waste disposal.

CYNTHIA L. PAULSON is manager of watershed services for Brown and Caldwell, an environmental engineering consulting firm in Denver, Colorado. She received a B.A. from Whitman College (political and environmental science), an M.S. from Colorado State University, and a Ph.D. from the University of Colorado (environmental engineering, 1987 and 1993). Dr. Paulson's work has focused on watershed and water quality planning and assessment, including evaluation of impacts on the physical, chemical, and biological integrity of surface waters and appropriate mitigation programs.

STUART S. SCHWARTZ is an associate hydrologic engineer with the Hydrologic Research Center, San Diego, California, where he leads the Center's technology transfer program. He received his B.S. and M.S. in biology-geology from the University of Rochester, and Ph.D. in systems analysis from the Johns Hopkins University. Before joining the Hydrologic Research Center, Dr. Schwartz was director of the Section for Cooperative Water Supply Operations on the Potomac (CO-OP) at the Interstate Commission on the Potomac River Basin. His research and professional interests focus on the application of systems analysis and multiobjective optimization in risk-based water resource management.

LEONARD SHABMAN received a Ph.D. in agricultural economics in 1972 from Cornell University. He is a professor at Virginia Polytechnic Institute and State University, Department of Agricultural and Applied Economics, and is director of the Virginia Water Resources Research Center. Dr. Shabman has conducted economic research over a wide range of topics in natural resource and environmental policy, with emphasis in six general areas: coastal resources management; planning, investment, and financing of water resource development; flood hazard management; federal and state water planning; water quality management; and fisheries management.

KAY D. THOMPSON is assistant professor at Washington University, Department of Civil Engineering. Her research is to investigate properties of subsurface materials for ground water studies, develop methods for subsurface characterization, assess the risks of hydrologic dam failure, and consult on minimizing environmental impacts during development. Dr. Thompson received a B.S. in civil engineering and operations research in 1987 from Princeton University, an M.S. in 1990 from Cornell University, and a Ph.D. in 1994 in civil and environmental engineering from the Massachusetts Institute of Technology.

DAVID A. WOOLHISER received his Ph.D. in civil engineering, with minors in meteorology and geophysics, from the University of Wisconsin in 1962. Dr. Woolhiser retired from the U.S. Department of Agriculture's Agricultural Research Service in 1991 after a 30-year career and is currently a faculty affiliate in civil engineering at Colorado State University and a hydrologist in Fort Collins, Colorado. He is known for his work on the hydrology and hydrometeorology of arid and semiarid rangelands, simulation of hydrologic systems, numerical modeling of surface runoff, erosion and chemical transport, and probabilistic models of rainfall and runoff. He is a member of the National Academy of Engineering.